THE SPECTROSCOPY OF
MOLECULAR IONS

THE SPECTROSCOPY OF MOLECULAR IONS

PROCEEDINGS OF
A ROYAL SOCIETY DISCUSSION MEETING
HELD ON 25 AND 26 MARCH 1987

ORGANIZED AND EDITED BY
A. CARRINGTON, F.R.S., AND B. A. THRUSH, F.R.S.

LONDON
THE ROYAL SOCIETY
1988

Printed in Great Britain for the Royal Society
by the
University Press, Cambridge

ISBN 0 85403 340 8

First published in *Philosophical Transactions of the Royal Society of London*,
series A, volume 324 (no. 1578), pages 73–294

Copyright

British Library Cataloguing in Publication Data

The Spectroscopy of molecular ions:
   proceedings of a Royal Society discussion
   meeting held on 25 and 26 March 1987.
   1. Ions   2. Molecular structure
   3. Spectrum analysis
   I. Carrington, A.   II. Thrush, B. A.
   III. Royal Society
   541.3′72   QD561

ISBN 0-85403-340-8

Published by the Royal Society
6 Carlton House Terrace, London SW1Y 5AG

CONTENTS

CONTENTS

Phil. Trans. R. Soc. Lond. A **324**, 75–80 (1988)

Printed in Great Britain

Directions in molecular-ion spectroscopy

By W. H. Wing

Department of Physics, University of Arizona, Tucson, Arizona 85721, U.S.A.

Molecular ions, once regarded as little more than curiosities, are now recognized as major (even dominant) chemical agents in extraterrestrial and terrestrial environments in which ionization occurs. Spectroscopic analysis offers the most comprehensive means of understanding their structure and following their behaviour. The field of charged-molecule spectroscopy is quite young but is growing rapidly. Some reasons for the rapid growth, particularly interesting situations involving molecular ions, and possibilities and problems for the future of the field are briefly discussed.

STATE OF THE FIELD

It is very satisfying for one who has been working for many years with those curiosities known as molecular ions to be able to attend a meeting devoted to that genus exclusively. Although the discipline of molecular spectroscopy is many decades old, interest in charged molecules has arisen much more recently. The rapid growth of the field is dramatically highlighted by comparing two comprehensive compilations of data, separated in time by about three decades, on diatomic molecules alone: Herzberg's volume (1950) and Huber & Herzberg's volume (1979) (table 1).

TABLE 1. NUMBERS OF KNOWN DIATOMIC MOLECULES

charge state...	0	+	−	2+	2−
1950	332	29	—	—	—
1979	786	141	40	6	—

In those 29 years, entries for neutral diatomics grew by a healthy 137%, from 332 to 786 molecules. Cations, although less numerous, showed nearly a five-fold increase, from 29 to 141 entries. Although no anions were known at all in 1950, 40 had been found by 1979. Even six entries of doubly charged cations appeared, nearly all based on theoretical work; no entries were made for doubly charged anions. The extent and quality of the data increased dramatically as well. And dramatic growth occurred also in the data for polyatomic neutral and charged molecules.

Among the most remarkable features of this explosion of knowledge is the way in which it has given substance to heretofore shadowy concepts in bulk chemistry. The hydronium ion H_3O^+, for example, was introduced to many of us in our first course in chemistry almost as a fictive device, to explain solubility and pH. Now molecular-ion spectroscopists have successfully isolated it *in vacuo*, have learned how to manipulate it, and have examined its inner workings in enormous detail; in short, they have made it very real indeed.

Since 1979 the growth in the field of molecular-ion spectroscopy has continued, as the very existence of this Discussion Meeting demonstrates. I dare say that a substantial proportion of the species that will be discussed here today and tomorrow were spectroscopically unknown in

[1]

1979. We are fortunate that many of those whose efforts have developed molecular-ion spectroscopy as a distinct subdiscipline are present. We may expect, therefore, a very stimulating two days.

Why, then, has all this activity taken place? What makes molecular ions so interesting? And what problems and opportunities may the field face in its future development? I should like to introduce our subject by offering very briefly some answers, of necessity incomplete and biased by my own interests, to these three questions.

The technology

The most direct reason for the rapid growth is surely the many advances in spectroscopic technology. Ion traps, ion beams, jet-expansion beams, cooled discharges, and other sources especially tailored to the controlled production of molecular ions have been extensively developed. Exotic varieties of lasers and other intense coherent sources in the infrared through to the submillimetre electromagnetic spectrum have become routine pieces of laboratory apparatus. Passive instruments such as the Fourier-transform and scanning Fabry–Perot interferometers have become reliable tools as well. Low-cost computers to control the equipment, take the data, and perform the analyses clutter laboratory benches. And major improvements in radiation detector sensitivity and spectral range have been made. With these new tools have come many ingenious methods for exploiting them, several of which we will hear about in the lectures to follow. The importance of advanced technology to molecular-ion spectroscopy can perhaps best be judged from the observation that the spectra of most neutral molecules were first detected by traditional grating or prism spectrometers, whereas those of most molecular ions were only found once less conventional methods could be applied.

Some interesting cases

What makes the field interesting is perhaps the most important question of the three. To the scientist–explorer, the first answer must be that it stirs the imagination. The simplest molecule of all, H_2^+, is a molecular ion. Its special role in the development of quantum theory is well known. The Schrödinger partial differential equation for its electronic motion is one of those few that can be separated into ordinary differential equations, and thus solved analytically, as was first done in 1927 (Burrau 1927).

The first molecule was also a molecular ion, as was the second molecule. Imagine the epoch at which the promordial nuclei have recently condensed and the radiation temperature is still several thousands of kelvins. The young Universe is a plasma sea composed mainly of photons, electrons, protons (H^+), and α-particles (He^{2+}). As expansion proceeds and the plasma cools, first He^+, then neutral He, then neutral H condense out (Lepp & Schull 1984). Once He appears, a burst of HeH^+ is formed from He and H^+ by radiative association and inverse rotational predissociation (Roberge & Dalgarno 1982). The rate is relatively small, however, at about 10^{-20} cm^2 s^{-1}. When H appears, a much larger burst of H_2^+ is formed by radiative association of H and H^+ (rate about 10^{-16} cm^2 s^{-1}).

The rest of the reason that any field is interesting must be the connections it makes, and the insight it has to offer, to other subjects. Let us briefly continue to follow, then, some connections molecular ions have to the other large-scale processes in the Universe.

When H_2^+ appears, chemistry begins (Dalgarno & Lepp 1987). Reactions with He produce more HeH^+, and those with H produce H_2. The H_2 destroys the primordial molecular ions, creating others, and a complex matrix of reactions ensues. The dominant processes become similar to those we see in the interstellar medium today.

This molecular-ion-driven chemistry strongly affected early molecular-cloud cooling and linkage to large-scale magnetic fields. Hence it affected the causal chain that begins with gravitational collapse rates and leads through the sizes of stars formed to the onset of nucleo-synthesis and the abundances of nuclides. It may have affected the scale of galaxy formation as well. In later, gentler times, molecular-ion catalysts strongly drive the rich chemistry of those clouds in which complex organic molecules are created (van Dishoeck & Black 1986). Did life originate in such a place? One need not speculate to realize how different our life would be today had not these several early manifestations of our subject presented themselves.

The special feature of molecular ions that gives the genus its very identity and positions them so importantly in processes such as these is, of course, the net non-zero charge. The charge interacts strongly with both nearby matter and the electromagnetic radiation field. At such long distances that a neutral molecule would ignore its neighbours, a molecular ion still polarizes surrounding material and thus, like any charged particle, tries to stick to whatever it encounters. At short distances it becomes a reactive chemical substance, and tends to transform whatever it sticks to. Thus molecular ions will dominate the chemistry of any medium in which they are formed in sufficient numbers, and the concentrations required are surprisingly small.

The ubiquitous earthly example of such a medium is surely an electrolyte. Every one of us is the possessor of such a substance; in fact, it is safe to assume that we all have several of them with us today. Although electrolytes have long been the province of the wet-laboratory methods of inorganic chemistry, the recent contributions of molecular-ion spectroscopy in elucidating structural geometry, molecular binding energies, and reaction channels have been substantial. I have already alluded to the hydronium ion and its role in electrolytes (and pedagogy). There is a curious irony here: the strong interactions of a molecular ion with its surroundings in condensed matter blur and distort the fine details of its spectrum. Nevertheless, without the ability to examine those details at the high resolution of gas-phase spectroscopy, much of our understanding of molecular structure would be lost. Thus there need be no tension between gas-phase and condensed-matter research, as sometimes exists. On the contrary, they go hand in hand.

For many of us here, space environments such as interstellar clouds, cometary atmospheres, and planetary outer atmospheres are the pre-eminent stages upon which molecular-ion processes play. Yet they also form copiously in plasma environments of possibly lesser inspirational value but certainly greater economic value to humanity: in flames and other combustion environments, in electrical discharges, and in controlled nuclear-fusion plasmas. Laser-based atomic spectroscopy is becoming a very useful tool for probing these environments, especially in understanding combustion processes. Because of the much higher complexity of the spectra, molecular-ion spectroscopy of these regions is less fully developed. But it seems likely to play an increasing role in the future.

The list of special molecular ions and their distinctive roles is long indeed. Yet one final addition may be of interest because of its novelty. The muonic hydrogen molecular ion is an exotic molecule whose valence electron has been replaced by a muon, which is some 200 times

[3]

more massive. The shorter de Broglie wavelength of this heavier lepton binds the molecule so closely together that, in its deuterated forms, the nuclei can actually undergo fusion. The muon is then released, and can repeat the process 150 or more times, as a kind of nuclear catalyst. The process was first noticed in cosmic muon collisions with photographic emulsions in the 1940s (Lattes *et al.* 1947) and quickly explained (Frank 1947; Sakharov 1948). It was rediscovered a decade later in bubble-chamber reactions by Alvarez and collaborators (1957). Recently, the discovery of a resonance in the reaction forming the deuterium–tritium isotope of this molecule (Thomsen 1987) has raised the possibility that it might be developed into a practical source of fusion power. Interest has revived, and an active international research community has developed.

PROBLEMS AND OPPORTUNITIES

Now very briefly to the third question: what may the future hold for molecular-ion spectroscopy, and what problems must we overcome? The speakers at this Meeting will answer the question with respect to their own topics in much more detail and with much greater precision than I could hope to, but perhaps a few speculative suggestions will not be out of place.

When a field is young, and new vistas have just opened themselves to us, we may happily explore them for nothing more than the pleasure that activity gives us. As the subject matures, however, the bridges that we can build from it to other knowledge become more and more necessary. Therefore we must ask what information those bridges require for their sturdiest construction. The single most critical bridge is that between experiment and underlying theory. It must be buttressed whenever possible.

One question that will become increasingly important in the future is how best to characterize reactions in which molecular ions participate. This is not entirely a spectroscopic question; yet spectroscopy is such a wonderful tool that it can unquestionably be of great service. An unmet need of theoretical astrophysicists, for example, is for the details of dissociative electron recombination reactions with polyatomic ions such as H_3O^+ and H_3^+. Data for other species are needed as well, and the need can only increase as our understanding of interstellar cloud composition grows.

In a very different domain, molecules have yet to take their place in the repertory of systems that can be used to test fundamental physical laws. One unique advantage that they possess for this purpose is that they hold two or more atomic nuclei in permanent close proximity. Surprisingly little is known about the fringes of nuclear force fields: high-speed nuclear collisions are simply too blunt an instrument to perform the delicate probing that is required. Suggestions have been made, for example, of van-der-Waals-like long-range nuclear forces resulting from exchanges of two or more massless particles such as gluons, very much as the ordinary van der Waals force between molecules can be viewed as arising from the exchange of pairs of photons (Feinberg & Sucher 1979; Coon 1986). Yukawa-like long-range forces may arise from exchange of massive particles, and significant forces may also arise from short-range modifications of the gravitational interaction. At present, however, the existence of all is purely speculative. It is interesting to note that the present best upper bound for the strengths of long-range forces that might arise from the exchange of up to four massless gluons, the so-called tensor long-range forces, comes from the agreement between theory and experiment for the hyperfine structure of the hydrogen molecule (Feinberg & Sucher 1979). Suitably designed experiments on the hydrogen molecular ion might provide a more stringent limit still, because of this molecule's simpler theoretical structure problem.

A broad difficulty in the interpretation of molecular spectroscopic data that is particularly keenly felt for certain small molecular ions is that the ability of theoretical quantum chemistry to predict has fallen far behind the ability of experimental spectroscopy to measure. This is surely the reason for the dearth of molecular experiments among tests of fundamental laws, despite their undeniable advantages, such as high spectroscopic Q.

It has become popular to believe that computational simulation of a phenomenon is essentially equivalent to studying the phenomenon itself. And indeed that is sometimes true, as we see from the recent dramatic shortening of product design cycles in many industries. In molecular physics as well, very substantial achievements have been made, particularly in the simulation of the overall geometry and structure of relatively complex organic molecules such as drugs. But on the whole, what is possible computationally still falls far short of what is required, and the gap with experiment is not closing. At the state of the art in molecular modelling, results for some situations can be achieved only by replacing quantum mechanics entirely by classical statistical-mechanical treatments whose connections to the physics can be remote at best (Beveridge & Jorgensen 1986). When quantum mechanics or indeed any dynamical theory is used, it is often severely limited. It was little more than a decade ago that internal motions were even recognized as an important feature of the structure of large molecules (Karplus 1986). Treatments of small molecules, although generally quantum-mechanical, have also been significantly unrealistic until recently. Of the hundreds of theoretical treatments of the hydrogen molecular ion from 1927 until 1976, when vibration–rotation spectra were observed (Wing *et al.* 1976), only a tiny handful went beyond the clamped-nucleus Born–Oppenheimer approximation, or considered corrections due to special relativity or vacuum polarization. Although this situation has been much improved, most other molecules are still treated at the Born–Oppenheimer level today, for the reason that their greatest source of computational error is the non-relativistic multielectron problem. Fresh theoretical approaches are badly needed.

Every major laboratory and every physical-chemistry journal contains reams of spectroscopic data that could not have been produced by present-day computer modelling, even though basic molecular quantum theory has been understood and accepted as valid for 60 years. The opportunities for improved analysis and insight seem considerable. Yet most experimentalists, for lack of something better, simply fit their results to a Dunham-like expansion about equilibrium (or do no analysis at all) and go to the next molecule. Although computer power and availability seemingly double every year or two, few theorists have yet attempted to match, or even approach, spectroscopic accuracy except for one-electron and two-electron molecules. Fewer still are attempting to solve structure and dynamics problems in the régimes of high excitation, far from equilibrium, where chaotic motion may exist (Berblinger *et al.* 1987), and yet where important reaction processess often occur.

A part of the problem is no doubt the tradition among many scientifically advanced nations of under-supporting theorists. Part may also be the natural time lag between the dissemination of a new technology and the achievement of practical results using it. Yet there may be important quantum theoretical problems that remain uncomputable in principle, by present methods at least, despite any foreseeable advance in computer technology, because their solution time will still exceed the lifetime of the computer, or the research grant, or the researcher, or the Universe, or some other natural barrier. May some seemingly finite problem, such as a truly first-principles solution of the Schrödinger problem for a single diatomic molecule composed of heavy atoms, be of such scope?

END AND BEGINNING

The answers to such questions are far from clear. Yet it is a sign of the progress of our field, I believe, that it is beginning to make sense to ask them. What does seem clear is that major advances will require new ideas for how and what to compute, how and what to measure and, most importantly, what to do with the results. The best last words on this subject seem to have been said by the numerical analyst R. W. Hamming (1973): 'The purpose of computing is insight, not numbers.' This applies, of course, to measurement as well.

Conversations with J. H. Black on astrophysics and S. A. Coon on nuclear forces materially improved this paper. The support of the Atomic–Molecular, and Plasma Physics Division of the U.S. National Science Foundation through a series of grants over the past 15 years is gratefully acknowledged.

REFERENCES

Alvarez, L. W. et al. 1957 Catalysis of nuclear reactions by μ-mesons. Phys. Rev. 105, 1127–1128.

Berblinger, M., Pollak, E. & Schlier, Ch. 1987 Bound states embedded in the continuum of H_3^+. (In the press.)

Beveridge, D. L. & Jorgensen, W. L. 1986 Preface in Computer simulations of chemical and biomolecular systems (ed. D. L. Beveridge & W. L. Jorgensen). Ann. N.Y. Acad. Sci., vol. 482.

Burrau, Ø. 1927 K. danske Vidensk. Selsk. Skr. 7, 14.

Coon, S. A. 1986 Meson theoretical models of three-nucleon forces. In Few body systems (suppl. 1) (ed. C. Ciofi degli Atti et al.), pp. 41–53. Vienna: Springer Verlag.

Dalgarno, A. & Lepp, S. 1987 Chemistry in the early universe. In Astrochemistry (ed. M. S. Vardya & S. P. Tarafdar), pp. 109–120. Dordrecht: D. Reidel.

Feinberg, G. & Sucher, J. 1979 Is there a strong van der Waals force between hadrons? Phys. Rev. D 20, 1717–1735.

Frank, F. C. 1947 Hypothetical alternative energy sources for the 'second meson' events. Nature, Lond. 160, 525–527.

Hamming, R. W. 1973 Numerical methods for scientists and engineers, 2nd ed. New York: McGraw-Hill.

Herzberg, G. 1950 Molecular spectra and molecular structure I. Spectra of diatomic molecules, 2nd ed. New York: Van Nostrand Reinhold.

Huber, K. P. & Herzberg, G. 1979 Molecular spectra and molecular structure IV. Constants of diatomic molecules. New York: Van Nostrand Reinhold.

Karplus, M. 1986 Molecular dynamics: application to proteins. In Computer simulations of chemical and biomolecular systems (ed. D. L. Beveridge & W. L. Jorgensen). (Ann. N.Y. Acad. Sci. vol. 482, pp. 255–266.)

Lepp, S. & Shull, J. M. 1984 Molecules in the early universe. Astrophys. J. 280, 465–469.

Lattes, C. M. G., Occhialini, G. P. S. & Powell, C. F. 1947 Observations on the tracks of slow mesons in photographic emulsions. Nature, Lond. 160, 453–456 and 486–492.

Roberge, W. & Dalgarno, A. 1982 The formation and destruction of HeH^+ in astrophysical plasmas. Astrophys. J. 255, 489–496.

Sakharov, A. D. 1948 Report, P. N. Lebedev Physics Institute, Academy of Sciences U.S.S.R., Moscow.

Thomsen, D. E. 1987 Keep cool with cold nuclear fusion. Science, Wash. 131, 133.

van Dishoeck, E. F. & Black, J. H. 1986 Comprehensive models of diffuse interstellar clouds: physical conditions and molecular abundances. Astrophys. J. 62 (suppl.), 109–145.

Wing, W. H., Ruff, G. A., Lamb, W. E. Jr & Spezeski, J. J. 1976 Observation of the infrared spectrum of the hydrogen molecular ion HD^+. Phys. Rev. Lett. 36, 1488–1491.

Phil. Trans. R. Soc. Lond. A **324**, 81–95 (1988)

Printed in Great Britain

Infrared spectroscopy of carbo-ions

By T. OKA, F.R.S.

*Department of Chemistry and Department of Astronomy and Astrophysics, The
University of Chicago, Chicago, Illinois* **60637,** *U.S.A.*

The recent development of high-resolution, high-sensitivity laser infrared spectroscopy has enabled us to study many fundamental molecular ions for which there had previously been no spectroscopic studies. We summarize our recent work here on carbo-ions, CH_3^+, C_2^-, $C_2H_2^+$ and $C_2H_3^+$.

1. INTRODUCTION

It is well known that the development of science occurs stepwise rather than continuously. In the area of molecular spectroscopy, the burst of activity in microwave spectroscopy after World War II, the electronic spectroscopy of free radicals in the 1950s and 1960s, the discoveries of many molecular lasers in the 1960s, and the radio-astronomical discoveries of interstellar molecules in the late 1960s and 1970s all occurred in this mode. I think we are now experiencing such a period in the field of molecular-ion spectroscopy. This Royal Society Meeting is very timely.

My venture in molecular-ion spectroscopy started with H_3^+, the simplest stable polyatomic system. Figure 1 shows the first 15 lines that were observed nearly seven years ago (Oka 1980). Since then a great many papers have been published on this fundamental molecular ion and its isotopic species (see, for example, a review by Sears 1987). Two weeks ago I was greatly excited by the discovery of the millimetre-wave spectrum of ArH_3^+ and ArD_3^+ complexes reported by Bogey *et al.* (1987). Last week I was also very surprised to hear from J. B. A.

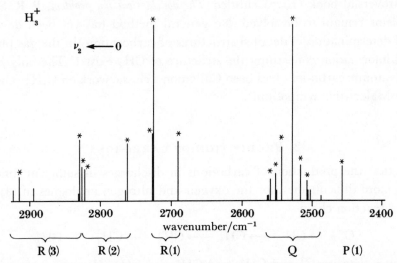

FIGURE 1. First 15 observed lines of the ν_2-band of H_3^+ (Oka 1980). Note that there is no obvious symmetry or regularity in the spectrum, although H_3^+ is a well-bound equilateral triangle. The analysis of the spectrum is by Watson (Watson 1984; Watson *et al.* 1984)

Mitchell that Smith & Adams (see Smith, this symposium) have now found that the electronic recombination rate constant of H_3^+ is less than 10^{-11} cm^3 s^{-1}, nearly that of atomic recombination. I look forward to hearing more about these most recent developments at this Meeting.

The H_3^+ ions are formed in a laboratory hydrogen discharge as well as in molecular clouds through the ion–molecule reaction

$$H_2^+ + H_2 \longrightarrow H_3^+ + H, \tag{1}$$

which will be formally written as a hydrogen-extraction reaction

$$H_2^+ \xrightarrow[-H]{H_2} H_3^+. \tag{2}$$

Likewise, protonated water H_3O^+ and protonated ammonia NH_4^+ are formed through the chains of hydrogen-extraction reactions

$$O^+ \xrightarrow[-H]{H_2} OH^+ \xrightarrow[-H]{H_2} H_2O^+ \xrightarrow[-H]{H_2} H_3O^+ \tag{3}$$

and

$$N^+ \xrightarrow[-H]{H_2} NH^+ \xrightarrow[-H]{H_2-} NH_2^+ \xrightarrow[-H]{H_2} NH_3^+ \xrightarrow[-H]{H_2} NH_4^+. \tag{4}$$

The reactions in (3) are all exothermic, with large Langevin cross sections. The same is true for the reactions in (4) except for the last step, which has a much smaller cross section. All the molecular ions in (3) and (4) have now been observed through their infrared spectra: OH$^+$ (Crofton *et al.* 1985); H$_2$O$^+$ (Dinelli *et al.* 1987); H$_3$O$^+$ (Begemann *et al.* 1983); NH$^+$ (Amano, this symposium); NH$_2^+$ (Rehfuss *et al.* 1987); NH$_3^+$ (Bawendi *et al.* 1987); and NH$_4^+$ (Crofton & Oka 1983).

Having studied some of these ions, it was natural to proceed to carbo-ions (here defined as molecular ions composed of only carbon and hydrogen atoms); we started searching for protonated methane CH_5^+ in the spring of 1983. The polyatomic hydrocarbon cations CH_n^+ ($n = 2$–5) and $C_2H_n^+$ ($n = 1$–7) are fundamental ionic species that play important roles in chemical kinetics in the laboratory and in space, yet there had previously been no spectroscopic studies of these in gaseous phase in any spectral region. In the concluding section of Herbert Brown's controversial book (1977) entitled *The nonclassical ion problem*, P. R. Schleyer notes: 'Major problems remain to be solved. No general method has yet been developed for the experimental determination of detailed structures of carbocations in the gas phase. No direct experimental information concerning the structure of CH_3^+ exists!' The only high-resolution studies of polyatomic carbo-ions had been Callomon's classic work on $C_4H_2^+$ (Callomon 1956), and work by Maier (this symposium).

2. Production of carbo-ions

We found that the production of carbo-ions in discharges in sufficient amount for spectroscopy was more difficult than for the oxygen and nitrogen analogues in (3) and (4). The corresponding reaction chain

$$C^+ \longrightarrow CH^+ \longrightarrow CH_2^+ \longrightarrow CH_3^+ \longrightarrow CH_4^+ \longrightarrow CH_5^+ \tag{5}$$

has two hangups between C$^+$ and CH$^+$, and CH$_3^+$ and CH$_4^+$ (Huntress 1977). The diatomic-carbon reaction chain is

$$C_2^+ \longrightarrow C_2H^+ \longrightarrow C_2H_2^+ \longrightarrow C_2H_3^+ \tag{6}$$

but does not extend further. The last reaction goes only for vibrationally excited $C_2H_2^+$ (Huntress 1977). The major difficulty in the production of carbo-ions results from the efficient polymerization of carbo-ions. Almost all reactions between carbo-ions and neutral hydro-carbons such as $CH_3^+ + CH_4(C_2H_4, C_2H_6$, etc.) have large Langevin cross sections and produce polymeric hydrocarbons, eventually leading to a large amount of soot deposited on the wall of a discharge tube. The best method to cope with this difficulty is to simply dilute the discharge gas mixture with a large amount of He. Helium also increases the efficiency of ion production through Penning ionization and, because of its high ionization potential, tends to increase the electron temperature in the plasma, thus fragmenting the larger carbo-ions. It is this use of He developed by my student Mark Crofton that has made possible the spectroscopy not only of carbo-ions but also of OH^+, H_2O^+, NH_2^+, NH_3^+, etc. We thus use a discharge gas mixture of $He:H_2:CH_4$ of $ca.$ 700:20:1 with a total pressure of $ca.$ 7 Torr† for the production of CH_3^+ and the same mixture with C_2H_2 for the production of $C_2H_3^+$. For a liquid-nitrogen-cooled discharge we used the former gas mixture even for $C_2H_3^+$ because C_2H_2 was frozen.

The relative abundance of different carbocations in (5) and (6) is very sensitive to the ratio of H_2 to hydrocarbon. The $C_2H_2^+$ lines disappear almost completely with an increased amount of H_2 in the discharge, whereas the $C_2H_3^+$ lines are not much affected. This fact was very useful to discriminate between the spectra of these two species that appear intermixed in the same region with similar spacings.

For the spectroscopy of carbo-ions, various multiple-inlet–outlet discharge tubes have been used. We have used an air-cooled discharge tube (nicknamed 'spider'), a water-cooled tube ('tarantula') and a liquid-nitrogen tube ('black widow') depending on purposes. The water-cooled tube is schematically shown in figure 2. A fresh mixture of gases is introduced into the discharge through the eight inlet ports. Blue-green emission, which Herzberg witnessed when he visited our laboratory and ascribed to CH, was noticed in the region of discharge around each inlet, indicating that hydrocarbon reactions are taking place vigorously in the optical path.

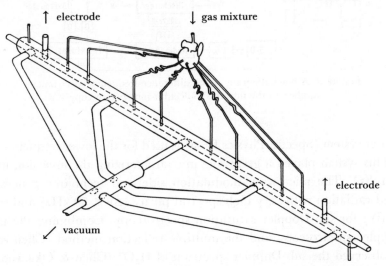

FIGURE 2. The water-cooled multiple-inlet–outlet discharge cell ('tarantula').

† 1 Torr ≈ 133.3 Pa.

[9]

3. SPECTROSCOPY

The success of infrared molecular-ion spectroscopy is caused by the great increase in sensitivity of infrared absorption spectroscopy brought about by laser spectroscopic techniques. The minimum detectable absorption is on the order of $\Delta I/I \approx 10^{-6}$–$10^{-7}$, indicating that this method is more sensitive than the traditional grating infrared spectroscopy by a factor of perhaps *ca.* 10^4–10^5.

We use three frequency-tunable laser infrared radiation sources for spectroscopy; they are a difference-frequency laser system, a diode-laser system, and a microwave-modulation sideband system on CO_2 laser lines. A diagram of the difference-frequency spectrometer is shown in figure 3. The radiation ν_A from a single-mode Ar ion laser (blue to green) and the radiation ν_D from a single-mode ring dye laser (yellow to red) are mixed in a temperature-controlled $LiNbO_3$ crystal to generate the infrared difference frequency $\nu_A - \nu_D$. This radiation source, initially developed by Pine (1974), generates continuously tunable infrared radiation over the infrared range of 2.2–4.2 μm with a power of *ca.* 100 μW.† This radiation source was indispensable for my first work on H_3^+ (Oka 1980) and is still the most powerful tool in our laboratory.

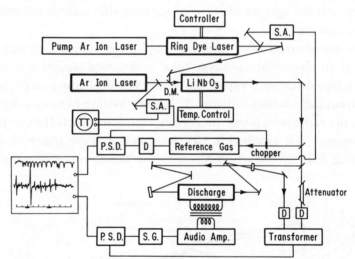

FIGURE 3. A block diagram of the difference-frequency spectrometer
applied to the infrared molecular-ion spectroscopy.

The diode-laser system (Spectra Physics LS-3) is used for the lower frequency range down to *ca.* 350 cm^{-1}. This system played a great role in characterizing the inversion motion of H_3O^+ (Liu & Oka 1985). The microwave-modulation sideband technique generates frequency-tunable infrared radiation with very high spectral purity ($\Delta\nu \leqslant 50$ kHz) and sufficiently high power (*ca.* 3 mW) for sub-Doppler saturation spectroscopy. Combining the ultra-high-resolution sub-Doppler spectroscopy and the multiple-reflection method (Chen *et al.* 1986), we have recently observed the sub-Doppler spectrum of H_3O^+ (Chen & Oka 1987).

To both discriminate the ion spectral lines from much stronger neutral absorption lines and

† 1 μW $= 10^{-6}$J s^{-1}.

to increase the sensitivity of ion spectroscopy, the powerful velocity-modulation technique developed by Gudeman *et al.* (1983) is used. This method is even more effective for carbo-ion spectroscopy because a great many neutral hydrocarbon molecules are present in the discharge and absorb infrared very strongly. We increase the optical path length by using a multiple-reflection mirror system. We use a unidirectional multiple path so that we can still use the velocity modulation.

4. OBSERVED SPECTRA

In March 1985 M. W. Crofton found many lines by using the air-cooled multiple-inlet discharge tube with the discharge gas mixture of He, H_2 and C_2H_2. Velocity modulation and chemistry indicated that the carriers of these spectral lines were hydrocarbon cations. About 1000 lines were observed in the region 3250–2990 cm^{-1}. These lines have been repeatedly studied under different chemical and plasma conditions. It has been revealed that these lines are caused by at least three carbocations and probably a few more. The beautiful spectral patterns of fundamental bands of the methyl cations CH_3^+, the acetylene ion $C_2H_2^+$, and the protonated acetylene $C_2H_3^+$ gradually emerged from this bush of lines.

(a) CH_3^+

A group of lines stood out as being caused by molecular ions with larger rotational constants because of their larger spacing of *ca.* 8 cm^{-1} and larger extension of R- and P-branches. Their larger linewidths also indicated smaller molecular masses. With CH_4 instead of C_2H_2, the intensities of these lines increased greatly. Initially we thought that the spectrum was caused by CH_5^+, for which *ab initio* calculation (Raghavachari *et al.* 1981) predicts $2B \approx 8$ cm^{-1}. It was soon found that the spectrum was too simple for that of CH_5^+ and that the observed spacing of 8 cm^{-1} corresponds to the spacing of 2 $(C-C\zeta)$ characteristic of the perpendicular band of an oblate top. More detailed analysis has established that this group of spectral lines correspond to the ν_3 fundamental band of CH_3^+.

Altogether 165 lines have been assigned up to the highest J of 16 and K of 16. A computer-generated stick spectrum of the observed vibration–rotation transitions is shown in figure 4. The spectrum is that of a typical perpendicular band for an oblate symmetric top. This appearance and the 2:1 intensity ratios for the $K = 3n$ and $K = 3n \pm 1$ transitions point to an equilateral triangle equilibrium structure. The absence of $J =$ even lines in the $^rQ_0$ branch reflects the planarity of the molecule and a totally symmetric A_1' electronic state. The D_{3h} geometrical structure of this ion had been expected from Herzberg's Rydberg spectrum of CH_3 (Herzberg 1961) and from *ab initio* theory. We have also observed and analysed the ν_3-band of $^{13}CH_3^+$.

Molecular constants determined from the spectrum by M.-F. Jagod are listed in table 1. A preliminary paper has been published (Crofton *et al.* 1985). More details of the results will be published soon (Crofton *et al.* 1987a). We have taken this opportunity to work out vibration–rotation interactions for this type of molecule in detail.

The methyl cation CH_3^+ plays a crucial role in interstellar chemistry of both diffuse clouds (Black & Dalgarno 1977) and dense clouds (Smith, this symposium). A chemistry chart in a diffuse molecular cloud is shown in figure 5. Future detection of this species in interstellar space through its infrared spectrum is awaited.

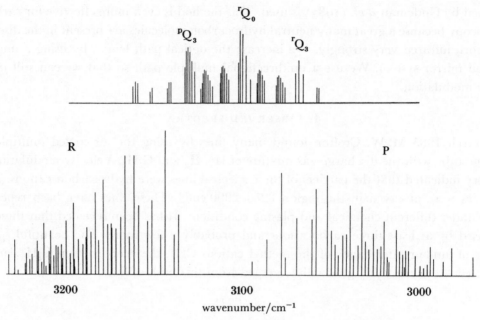

FIGURE 4. A computer-generated stick diagram of the observed ν_3 fundamental band of CH_3^+.

TABLE 1. DETERMINED MOLECULAR CONSTANTS OF CH_3^+ AND $^{13}CH_3^+$

(RECIPROCAL CENTIMETRES)

	CH_3^+	$^{13}CH_3^+$
$\nu_3 - C'\zeta_{33} - \frac{1}{4}\eta_K$	3107.8559 (53)	3095.2133 (97)
B'	9.27224 (44)	9.2727 (17)
$C'(1-\zeta_{33}) + \eta_K$	4.04591 (84)	4.0969 (22)
$C' - C'' + \frac{3}{2}\eta_K$	-0.04006 (27)	0.0405 (17)
D'_{JJ}	0.7104 (78) $\times 10^{-3}$	0.679 (51) $\times 10^{-3}$
D'_{JK}	-1.313 (21) $\times 10^{-3}$	-1.274 (99) $\times 10^{-3}$
$D'_{KK} - \frac{1}{4}\eta_K$	0.469 (16) $\times 10^{-3}$	0.481 (57) $\times 10^{-3}$
η_J	-0.643 (44) $\times 10^{-3}$	-0.60 (19) $\times 10^{-3}$
B''	9.36227 (31)	9.3621 (16)
D''_{JJ}	0.7482 (65) $\times 10^{-3}$	0.720 (50) $\times 10^{-3}$
D''_{JK}	-1.337 (16) $\times 10^{-3}$	-1.275 (99) $\times 10^{-3}$
$D''_{KK} - D'_{KK}$	0.019 (13) $\times 10^{-3}$	0.011 (61) $\times 10^{-3}$
$H'_{JJJ} = H''_{JJJ}$	0.106 (41) $\times 10^{-6}$	
$H'_{JJK} = H''_{JJK}$	-0.31 (11) $\times 10^{-6}$	
$H'_{JKK} = H''_{JKK}$	0.23 (11) $\times 10^{-6}$	
$H'_{KKK} = H''_{KKK}$	-0.16 (13) $\times 10^{-6}$	
q	-8.6 (10) $\times 10^{-3}$	-11.4 (10) $\times 10^{-3}$
q_J	-0.026 (15) $\times 10^{-3}$	
q_K	-1.40 (38) $\times 10^{-3}$	
β_2	-0.0367 (25) $\times 10^{-3}$	-0.0414 (71) $\times 10^{-3}$
h_3	-0.0112 (61) $\times 10^{-6}$	

(b) C_2^-

Infrared spectra of C_2^- were observed first in the 2.6 μm region by using the difference-frequency system and later in the 4.6 μm region with the diode-laser system. These spectra are different from other carbo–ion spectra in various ways: (a) they are electronic transitions $A^2\Pi_u \leftarrow X^2\Sigma_g$ instead of the usual vibrational transitions; (b) the carrier of this spectrum is

[12]

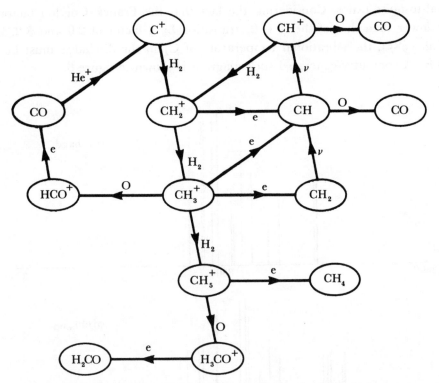

FIGURE 5. Interstellar chemistry diffuse molecular cloud after Dalgarno.

an anion; and (c) quantum mechanics of this ion had earlier been studied in relation to its electronic spectra. However, our study of this ion and thus the $^2\Pi$ state was essential in assigning the $C_2H_2^+$ spectrum discussed in the next section.

The first spectroscopic study of C_2^- is described in a classic paper by Herzberg & Lagerqvist (1968), in which they report a new system of $\Sigma-\Sigma$ bands in the region 4800–6000 Å† and point out the possibility that the bands are caused by C_2^-. Milligan & Jacox (1969) observed the same transition in inert gas matrices and reported that the band is very likely from C_2^- because its intensity increased with the presence of an electron donor. The final confirmation came when Lineberger & Patterson (1972) observed the two-photon photodetachment spectrum of C_2^-. These results characterised the $B^2\Sigma_u^+$ state and the $X^2\Sigma_g^+$ state with high accuracy. More recently, Mead *et al.* (1985) observed the B ← X transition with sub-Doppler resolution and by using the perturbations on the B state from the high vibrational states of the $A^2\Pi_u$ state, determined some molecular constants for the $A^2\Pi_u$ state. We used their constants, in particular the A ← X energy separation, as the guide for our search of the spectrum.

The infrared spectrum of C_2^- was found by B. Dinelli and M.-F. Jagod in January 1986. It was found that a mixture of a large amount of He (7 Torr) and a small amount of C_2H_2 (*ca.* 50 mTorr) was optimum for the most effective production of C_2^-. We estimate the C_2^- concentration in the discharge to be *ca.* 10^8 cm^{-3}. Although this is considerably less than the concentration of other carbo-ions, the spectrum is rather strong because the transition dipole moment is much larger. The vibrationally hot bands (1 ← 1) and (0 ← 1) were observed with

† 1 Å = 10^{-1} nm = 10^{-1} m.

good signal-to-noise ratios. Considering the fact that the Franck–Condon factors for these bands are lower than that of the $(0 \leftarrow 0)$ transition by a factor of 2.0 and 3.2, respectively (Mead *et al.* 1985), the vibrational temperature of C_2^- in the discharge must be very high: $T \gtrsim 5000$ K. A computer-generated stick diagram is shown in figure 6.

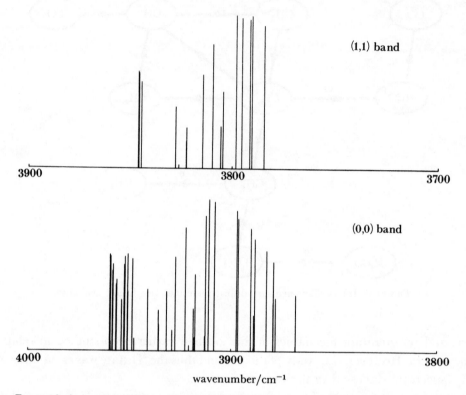

FIGURE 6. A computer-generated stick diagram of the observed $^2\Pi \leftarrow {}^2\Sigma_g^+$ transition of C_2^-.

Since the initial theoretical work by Hill & Van Vleck (1928), a great many papers have been published on the hamiltonian for the $^2\Pi$ state. We used the hamiltonian by Brown & Watson (Brown *et al.* 1979).

$$H = BN^2 - DN^4 + \tfrac{1}{2}[(\tilde{A} + \tilde{A}_D N^2), L_z S_z]_+ + \tfrac{1}{2}p(\Lambda_+^2 S_- N_- + \Lambda^2 S_+ N_+) - \tfrac{1}{2}q(\Lambda_+^2 N_- + \Lambda^2 N_+^2). \quad (7)$$

For the derivation of this formula and the definition of the parameters and operators, refer to the paper by Brown *et al.* (1979). For the ground $^2\Sigma_g^+$ state the usual hamiltonian

$$H = BN^2 - DN^4 + \gamma N \cdot S \qquad (8)$$

was used. The 104 observed lines for the three observed bands have been fitted simultaneously by a least-squares computer program. The parameters determined for the $A^2\Pi_u$ state are shown in table 2, where previously determined parameters are also shown. The agreement with the results of Mead *et al.* (1985) is remarkable considering the indirect nature of their determination.

[14]

TABLE 2. MOLECULAR CONSTANTS (RECIPROCAL CENTIMETRES) OF C_2^- IN THE $A^2\Pi_u$ STATE

	this work	Mead et al. (1985)
T_0	3928.660 (17)	4002 (91)
ν_0	1644.803 (51)	1637 (10)
A_0	−24.989 (62)	−24 (1)
A_e	−25.032 (56)	
α_A	−0.084 (69)	
B_0	1.634 94 (26)	1.622
B_e	1.643 07 (51)	1.630 (5)
α_B	0.016 25 (72)	0.015 2
r_e	1.307 67 (20) Å	1.313

(c) $C_2H_2^+$

After our excursion to C_2^-, we went back to the complicated spectrum in the 3.3 μm region that we thought was all caused by $C_2H_3^+$. *Ab initio* calculations on this molecular ion have indicated that two structures are possible, the formaldehyde-type (classical) structure and the bridged (non-classical) structure, and that they are very close in energy (Weber *et al.* 1987; Raine & Schaefer 1984). In the classical structure, the C_2-axis for permutation of the two equivalent protons is along the *a*-axis, whereas in the bridged structure the axis is along the *b*-axis. The former will show *K*-doubling spectral lines with equal intensities whereas the latter will give a ratio of 3:1.

There was a series of doublet lines with equal intensities in the bush of the spectrum that initially led us to think $C_2H_3^+$ has the classical structure. However, after more careful analysis with combination differences together with the newly gained understanding of the $^2\Pi$ state, and after the chemical experiment discussed earlier, we realized that they were caused by the ν_3-band $^2\Pi \leftarrow {}^2\Pi$ transitions of the acetylene ion $C_2H_2^+$. Extremely weak Q-branch lines expected for a $\Pi \leftarrow \Pi$ transition played a crucial role in the final assignment. A preliminary paper has been published (Crofton *et al.* 1987 *b*). The molecular constants determined are listed in table 3. A computer-generated stick diagram of the observed transitions is shown in figure 7. Whereas low *J* transitions $J < 10$ fit well to the usual $^2\Pi$ hamiltonian, higher *J* transitions deviate markedly from the calculated spectrum. More detailed analysis is in progress. The present results confirm the earlier estimate by Hollas & Sutherlay (1971) that the C–C bond length increases by about 0.04–0.05 Å upon ionization of acetylene.

TABLE 3. MOLECULAR CONSTANTS (RECIPROCAL CENTIMETRES) OF $C_2H_2^+$ $^2\Pi_u$

$$\nu_3 = 3135.975\ (5)$$
$$B' = 1.098\,95\ (13) \qquad A' = -29.6\ (15)$$
$$B_0 = 1.104\,56\ (7) \qquad A_0 = -30.1\ (15)$$
$$\alpha_3 = 0.005\,61\ (15)$$
$$D_0 = D' = 1.44 \times 10^{-6}$$

(d) $C_2H_3^+$

After assigning many spectral lines of CH_3^+ and $C_2H_2^+$, we were left with more than 1000 lines in the 3200–3090 cm^{-1} region. It is possible that these lines are caused by more than one carbocation, but a major portion of them are believed to be from protonated acetylene

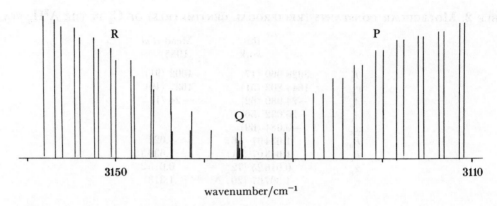

FIGURE 7. A computer-generated stick diagram of the ν_3-band $^2\Pi \leftarrow {}^2\Pi$ transition of $C_2H_2^+$.

$C_2H_3^+$. A computer-generated stick diagram of the central portion of the observed spectrum is shown in figure 8. The $C_2H_3^+$ lines are intermixed with CH_3^+ and $C_2H_2^+$ lines as shown in figure 9. In particular, both the band origins and the spacings of groups of lines of $C_2H_2^+$ and $C_2H_3^+$ are very close; this confused us at the initial stage of the assignment.

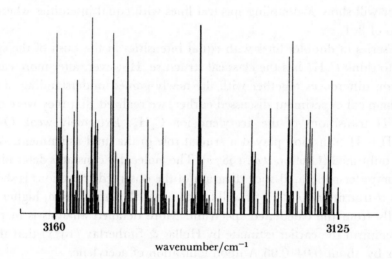

FIGURE 8. A stick diagram for the central portion of the observed $C_2H_3^+$ spectrum.

Although the intensities of the $C_2H_2^+$ lines depend critically on the hydrogen concentration, the $C_2H_3^+$ lines are produced in almost any discharge containing hydrocarbon, H_2 and He. We obtained similar spectra when we used C_2H_4, C_2H_6, C_2H_3Br, and CH_2CCH_2, as when we used C_2H_2. We produced the spectrum with D_2 instead of H_2 in various mixing ratios and ^{13}C-enriched HCCH. These results were not inconsistent with the carrier of the gas being $C_2H_3^+$.

Low J spectral lines assigned so far are listed in table 4. There are other lines that are assigned, but their assignment is not as definitive as those listed in table 4. These spectral lines are close to the asymmetric rotor pattern expected for the rigid non-classical structure. The 'staggering' behaviour of the $K = 1$ K-doubling lines clearly shows that the ion has the bridged structure.

[16]

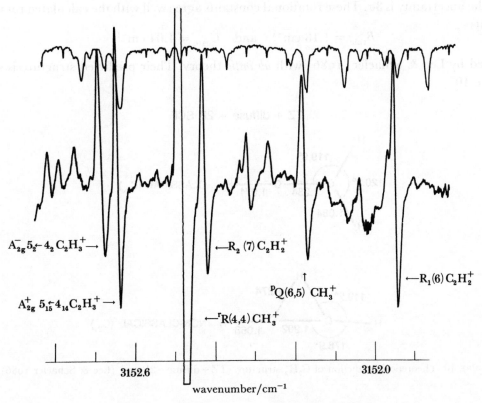

FIGURE 9. An example of carbo-ion spectral lines.

TABLE 4. TENTATIVE ASSIGNMENT (RECIPROCAL CENTIMETRES) OF THE LOW J LINES OF $C_2H_3^+$

$9_{28} \leftarrow 8_{27}$	A_{2g}^-	3161.160		$2_{11} \leftarrow 1_{10}$	A_{2g}^+	3146.708
$9_{19} \leftarrow 8_{18}$	A_{2g}^+	3161.042		$2_{02} \leftarrow 1_{01}$	A_{2g}^-	3146.633
$8_{17} \leftarrow 7_{16}$	A_{2g}^+	3159.511		$2_{02} \leftarrow 3_{03}$	A_{2g}^-	3135.690
$8_{08} \leftarrow 7_{07}$	A_{2g}^-	3159.316		$2_{11} \leftarrow 3_{12}$	A_{2g}^+	3135.540
$8_{26} \leftarrow 7_{25}$	A_{2g}^-	3159.041		$3_{13} \leftarrow 4_{14}$	A_{2g}^+	3133.453
$8_{18} \leftarrow 7_{17}$	E_g^-	3159.189		$4_{14} \leftarrow 5_{15}$	E_g^-	3131.292
$7_{16} \leftarrow 6_{15}$	E_g^-	3157.396		$4_{04} \leftarrow 5_{05}$	A_{2g}^-	3131.232
$7_{17} \leftarrow 6_{16}$	A_{2g}^+	3156.962		$4_{13} \leftarrow 5_{14}$	A_{2g}^+	3130.776
$7_{26} \leftarrow 6_{25}$	A_{2g}^-	3156.931		$5_{15} \leftarrow 6_{16}$	A_{2g}^+	3129.104
$6_{15} \leftarrow 5_{14}$	A_{2g}^+	3155.204		$5_{05} \leftarrow 6_{06}$	E_g^+	3128.870
$6_{24} \leftarrow 5_{23}$	A_{2g}^-	3154.756		$5_{24} \leftarrow 6_{25}$	A_{2g}^-	3128.636
$6_{16} \leftarrow 5_{15}$	E_g^-	3154.672		$5_{14} \leftarrow 6_{15}$	E_g^-	3128.406
$5_{14} \leftarrow 4_{13}$	E_g^-	3152.989		$6_{24} \leftarrow 7_{25}$	A_{2g}^-	3126.363
$5_{05} \leftarrow 4_{04}$	E_g^+	3152.897		$6_{15} \leftarrow 7_{16}$	A_{2g}^+	3126.150
$5_2 \leftarrow 4_2$	A_{2g}^-, E_g^+	3152.678		$7_{17} \leftarrow 8_{18}$	A_{2g}^+	3124.876
$5_{15} \leftarrow 4_{14}$	A_{2g}^-	3152.637		$7_{07} \leftarrow 8_{08}$	E_g^+	3124.534
$4_{04} \leftarrow 3_{03}$	A_{2g}^-	3150.908		$8_{08} \leftarrow 9_{09}$	A_{2g}^-	3122.168
$4_{13} \leftarrow 3_{12}$	A_{2g}^+	3150.889		$8_{26} \leftarrow 9_{27}$	A_{2g}^-	3121.776
$4_{14} \leftarrow 3_{13}$	E_g^-	3150.540		$8_{17} \leftarrow 9_{18}$	A_{2g}^+	3121.535
$3_{13} \leftarrow 2_{12}$	A_{2g}^+	3148.438				

If we take combination differences in the ground state, the 23 differences can be fitted fairly well to an asymmetric rotor model (standard deviation of the fit is *ca.* 0.02 cm^{-1}) with the following rotational constants:

$$B = 1.1412 \ (19) \ \text{cm}^{-1} \quad \text{and} \quad C = 1.0467 \ (19) \ \text{cm}^{-1},$$

where the uncertainty is 3σ. These rotational constants agree well with the calculated rotational constants

$$B_{\text{calc}} = 1.13 \text{ cm}^{-1} \quad \text{and} \quad C_{\text{calc}} = 1.04 \text{ cm}^{-1}$$

predicted by Lee & Schaefer (1986) with *ab initio* theory. Their predicted structure is shown in figure 10.

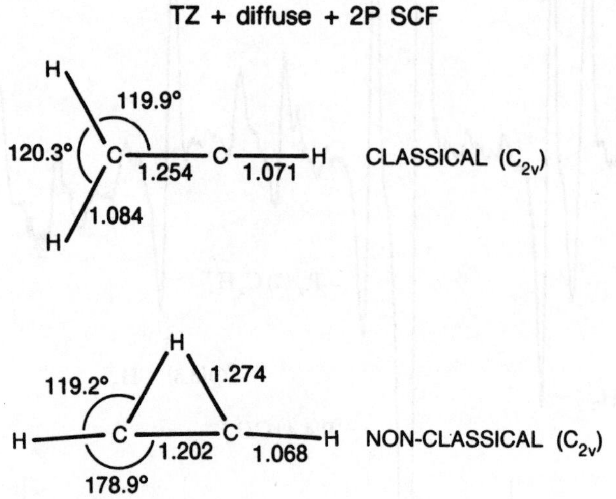

TZ + diffuse + 2P SCF

FIGURE 10. Theoretical prediction of $C_2H_3^+$ structure (TZ + diffuse + 2P scf) (Lee & Schaefer 1986).

Although these results establish the carrier of the spectrum to be $C_2H_3^+$, with the non-classical structure dominating (a conclusion earlier obtained by Kanter *et al.* (1986) from their Coulomb explosion experiment), there are various indications that the spectrum is not that of a usual well-behaved asymmetric rotor. They are as follows. (1) The standard deviation of fitting is larger than the normal case by a factor of 10. In particular, the lowest J transitions ($J = 0$, 1, 2) have large residuals. (2) The observed intensity ratios of the stronger lines to weaker lines of K-doublets (which have nearly equal transition matrix elements except for the spin weight) are closer to 2:1 rather than the 3:1 expected from the usual permutation of two protons. (3) Spectral lines with higher K values ($K \geqslant 3$) are more difficult to assign if we impose combination differences involving Q-branch lines. (4) For each group of lines, there are extra lines with significant intensities. These observations may be associated with the proton tunnelling expected from the theoretically calculated small energy difference between the non-classical and classical ions and the small barrier separating them (Weber *et al.* 1976).

Recently Hougen (1987) and Escribano & Bunker (1987) developed a theory in which the three protons in $C_2H_3^+$ are tunnelling through the barrier and rotating in the plane of $C_2H_3^+$. In such a model the three protons are all equivalent in contrast to the rigid C_{2v} model, in which only two protons are equivalent. The spin-weight ratio is then 2:1 as in NH_3 rather than 3:1 as in H_2. We should use for symmetry argument the full permutation–inversion group

$$S_3 \otimes S_2 \otimes E^* = G_{24},$$

which is isomorphic to D_{6h} ($E^* \equiv \{E, E^*\}$). A simple symmetry argument shows that the symmetry of the total wavefunction should be either A_{2g}^+ or A_{2g}^- and that of the spin wavefunction $4A_{1g}^+$ (for $I = \frac{3}{2}$) and $2E_g^+$ (for $I = \frac{1}{2}$). The correlation table (below) shows that the stronger components of the K doublet that have a spin weight of 3 are split into two lines with the weights of 2 and 1:

C_{2v}	$K_a K_c$	G_{24}
A_1	ee	E_g^+
A_2	oo	E_g^-
$3B_1$	oe	$2A_{2g}^+ + E_g^+$
$3B_2$	eo	$2A_{2g}^- + E_g^-$

Hougen (1987) calculated the magnitude of the splitting based on the internal-rotation hamiltonian used for the analysis of the microwave spectra of CH_3BH_2 and CH_3NO_2 (Wilson *et al.* 1955; Tannenbaum *et al.* 1956). The application of Hougen's programme to the analysis of the $C_2H_3^+$ is in progress.

5. CONCLUSIONS

Spectroscopy of simple carbo-ions has just begun. None of the work described in this paper has been completed yet. Many spectral lines have been observed but their identity is unknown at present. A great many other carbo-ions will be studied. We have already invested a considerable amount of time searching for the spectra of C_2H^-, $C_3H_3^+$ and CH_5^+. Their spectra and many others will be observed and characterized in the near future. The CH_5^+ spectrum will be particularly interesting. The permutation of five protons, S_5, will be involved. This group, which has a high degeneracy, has never been used in molecular spectroscopy. These results will provide fresh information for quantum chemistry, chemical kinetics, plasma physics and astrophysics. Above all, together with spectroscopy of other ions reported in this Meeting, they enrich chemistry at the most fundamental level.

This work has been supported by N.S.F. Grant PHY-84-08316.

REFERENCES

Bawendi, M. G., Rehfuss, B. D., Dinelli, B. M., Okumura, M. & Oka, T. 1987 In *Molecular Spectroscopy Symposium, Columbus, Ohio*, p. 51 (abstract).
Begemann, M. H., Gudeman, C. S., Pfaff, J. & Saykally, R. J. 1983 *Phys. Rev. Lett.* **51**, 554–557.
Black, J. H. & Dalgarno, A. 1977 *Astrophys. J. Suppl.* **34**, 405–423.
Bogey, M., Bolvin, H., Demuynck, C. & Destombes, J. L. 1987 *Phys. Rev. Lett.* **58**, 988–991.
Brown, H. C. 1977 *The Nonclassical Ion Problem*, Plenum Press, New York & London.
Brown, J. M., Colbourn, E. A., Watson, J. K. G. & Wayne, F. D. 1979 *J. molec. Spectrosc.* **74**, 294–318.
Callomon, J. H. 1956 *Can. J. Phys.* **34**, 1046–1074.
Chen, Y. T., Frye, J. M. & Oka, T. 1986 *J. opt. Soc. Am.* B 3, 935–939.
Chen, Y. T. & Oka, T. 1987 In *Molecular Spectroscopy Symposium, Columbus, Ohio*, p. 52 (abstract).
Crofton, M. W., Altman, R. S., Jagod, M.-F. & Oka, T. 1985 *J. phys. Chem.* **89**, 3614–3617.
Crofton, M. W. & Oka, T. 1983 *J. chem. Phys.* **79**, 3157–3158.
Crofton, M. W., Kreiner, W. A., Jagod, M.-F., Rehfuss, B. D. & Oka, T. 1985 *J. chem. Phys.* **83**, 3702–3703.
Crofton, M. W., Jagod, M.-F., Rehfuss, B. D., Kreiner, W. A. & Oka, T. 1987a. (In preparation.)
Crofton, M. W., Jagod, M.-F., Rehfuss, B. D. & Oka, T. 1987b *J. chem. Phys.* **86**, 3755–3756.
Dinelli, B., Crofton, M. W. & Oka, T. 1987 *J. molec. Spectrosc.* (In the press.)
Escribano, R. & Bunker, P. R. 1987 *J. molec. Spectrosc.* **122**, 325–340.
Gudeman, C. S., Begemann, M. H., Pfaff, J. & Saykally, R. J. 1983 *Phys. Rev. Lett.* **50**, 727–731.
Herzberg, G. 1961 *Proc. R. Soc. Lond.* A **262**, 291–317.
Herzberg, G. & Lagerqvist, A. 1968 *Can. J. Phys.* **46**, 2363–2373.
Hill, E. & Van Vleck, J. H. 1928 *Phys. Rev.* **32**, 250–272.
Hollas, J. M. & Sutherlay, T. A. 1971 *Molec. Phys.* **21**, 183–185.
Hougen, J. T. 1987 *J. molec. Spectrosc.* **123**, 197–227.
Huntress, W. T. Jr 1977 *Astrophys. J. Suppl.* **33**, 495–514.

Kanter, E. P., Vager, Z., Both, G. & Zajfman, D. 1986 *J. chem. Phys.* **85**, 7487–7488.

Lee, T. J. & Schaefer, H. F. III 1986 *J. chem. Phys.* **85**, 3437–3443.

Lineberger, W. C. & Patterson, T. A. 1972 *Chem. Phys. Lett.* **13**, 40–44.

Liu, D.-J. & Oka, T. 1985 *Phys. Rev. Lett.* **54**, 1787–1789.

Mead, R. D., Hefter, U., Schulz, P. A. & Lineberger, W. C. 1985 *J. chem. Phys.* **82**, 1723–1731.

Milligan, D. E. & Jacox, M. E. 1969 *J. chem. Phys.* **51**, 1952–1955.

Oka, T. 1980 *Phys. Rev. Lett.* **45**, 531–534.

Pine, A. S. 1974 *J. opt. Soc. Am.* **64**, 1683–1690.

Raghavachari, K., Whiteside, R. A., Pople, J. A. & Schleyer, P. V. R. 1981 *J. Am. chem. Soc.* **103**, 5649–5657.

Rehfuss, B. D., Dinelli, B. M., Okumura, M., Bawendi, M. G. & Oka, T. 1987 In *Molecular Spectroscopy Symposium, Columbus, Ohio*, p. 51 (abstract).

Raine, G. & Schaefer, M. F. II 1984 *J. chem. Phys.* **81**, 4034–4037.

Sears, T. J. 1987 *J. chem. Soc. Faraday Trans.* II **83**, 111–126.

Tannenbaum, E., Myers, R. J. & Gwinn, W. D. 1956 *J. Chem. Phys.* **25**, 42–47.

Watson, J. K. G. 1984 *J. molec. Spectrosc.* **103**, 350–363.

Watson, J. K. G., Foster, S. C., McKellar, A. R. W., Bernath, P., Amano, T., Pan, F. S., Crofton, M. W., Altman, R. S. & Oka, T. 1984 *Can. J. Phys.* **62**, 1875–1885.

Weber, J., Yoshimine, M. & McLean, A. D. 1976 *J. chem. Phys.* **64**, 4159–4164.

Wilson, E. B., Lin, C. C. & Lide, D. R. 1955 *J. chem. Phys.* **23**, 136–141.

Discussion

R. SAYKALLY (*Department of Chemistry, University of California, Berkeley, U.S.A.*). Regarding Professor Oka's tentative assignment of the vibrational spectrum of protonated acetylene to a model with the hydrogens tunnelling between the classical and non-classical forms with an apparent barrier near 600 cm^{-1}, it is a bit troublesome that the most recent energy-level *ab initio* calculations on this system indicate a large separation of the two isomers (*ca.* 3 kcal; 1 kcal = 4.18×10^3 J). Also the Argonne Coulomb explosion experiment provides evidence only for the non-classical structure. Can he reconcile these facts with his assignment?

T. OKA, F.R.S. We have not attempted to determine the barrier by fitting the spectrum. The barrier is probably much higher than initially predicted by Weber *et al.* (1976).

R. SAYKALLY. Has Professor Oka observed hot bands for either the methyl cation or the acetylene cation in his spectra? In the latter case, the complex Renner–Teller interactions may make identification very difficult. Also, what are the rotational temperatures observed for these two ions? Do these change substantially with conditions?

T. OKA, F.R.S. We have observed many extra lines for CH_3^+ and $C_2H_2^+$, but none of them have so far been definitively assigned to hot bands. The bending-excited hot bands of $C_2H_2^+$ will be very complicated by the Renner–Teller effect. We have not attempted to determine rotational temperatures, but typical values are 200–300 K for a liquid-N_2 cooled discharge, 500 K for a water-cooled discharge and 800–1000 K for an air-cooled discharge. We used an air-cooled discharge tube for observing high J, K lines of CH_3^+.

H.-J. FOTH (*University of Kaiserlauten, F.R.G.*). Could Professor Oka give more information about the last experiment of H_3O^+? By which techniques did he obtain sub-Doppler resolution. Is it a kind of saturation spectroscopy?

T. OKA, F.R.S. We used the inverse-Lamb dip technique for this observation. Mixing microwave radiation (ν_m) and CO_2 infrared laser radiation (ν_L) in a CdTe crystal, we obtain frequency-tunable infrared radiation in the 10 μm region with high spectral purity and sufficient power for saturation. We combined this source with a low-pressure (30 mTorr) hollow cathode discharge to observe the sub-Doppler spectrum.

E. HIROTA (*Institute for Molecular Science, Okazaki* 444, *Japan*). There are three 'C–H' stretching states in protonated acetylene. Which one is the upper state of the transition that Professor Oka assigned?

T. OKA, F.R.S. The observed excited state corresponds to the antisymmetric C–H stretching vibration of the bridged (non-classical) structure. The band origin is 3142 cm^{-1}.

Phil. Trans. R. Soc. Lond. A **234**, 97–108 (1987)

Printed in Great Britain

Velocity-modulation infrared laser spectroscopy of molecular anions

J. Owrutsky, N. Rosenbaum, L. Tack, M. Gruebele, M. Polak
and R. J. Saykally

Department of Chemistry, University of California, Berkeley, California 94720, U.S.A.

The velocity-modulation technique has been used with colour-centre lasers and diode lasers throughout the vibrational infrared region to measure high-resolution vibration–rotation spectra of fundamental anions.

Introduction

Although molecular anions are of fundamental significance in chemistry and in related disciplines, high-quality information on their structures, properties, and dynamics has been very difficult to obtain. Existing experimental knowledge of the structures and properties of negative ions comes principally from diffraction and low-resolution spectroscopy experiments carried out in condensed phases and from gas-phase electron detachment studies made in ion traps and ion beams. The latter techniques have been developed by Lineberger, Brauman and others into elegant and powerful tools for measuring electron affinities, energies and in some cases approximate geometries of negative ions and their associated neutral radicals. Nevertheless, because negative ions have not been amenable to absorption spectroscopy, there have been virtually no precise measurements of their molecular structures and potential functions. *Ab initio* theoretical calculations, which can be carried out with high accuracy for molecular cations, are more difficult for negatively charged molecules, and their reliability in this context is relatively untested.

Two complementary new experimental techniques for measuring high-resolution vibration–rotation spectra of negative ions have been developed in the last two years. Lineberger and his colleagues at the Joint Institute for Laboratory Astrophysics have perfected a laser electron detachment approach that they call 'vibrational autodetachment spectroscopy'; Neumark *et al.* (1985); in this technique a tunable infrared laser excites vibrational transitions in a beam of negative ions for which the electron binding energy is less than the energy of the laser photon. The excited vibrational state is thus 'quasibound' and autodetaches the extra electron on a time scale determined by the electron–nuclear coupling terms neglected in the Born–Oppenheimer approximation. This timescale varies over orders of magnitude with the quantum state of the anion, but is long enough to yield hyperfine-resolved vibration–rotation spectra, as well as important information on the dynamics of the autodetachment process. Neumark *et al.* (1985) have used this technique to carry out the first spectroscopic study of NH⁻, which has been extended in recent work by Miller & Farley (1987) and Al-Za'al *et al.* (1986, 1987). Similar experiments have been carried out with visible lasers for $C_2H_3O^-$ (acetaldehyde enolate), CH_2CN^-, and FeO^-, in which states having the additional electron bound to the permanent dipole moment of the neutral molecule were observed (Read *et al.* 1984; Lykke *et al.* 1984; Anderson *et al.* 1987). It thus appears that autodetachment spectroscopy will be applicable to a large class of negative ions.

The most stable molecular anions have filled valence shells and are often the conjugate bases of common simple acids; these are the so called 'classical anions'. Some of the more ubiquitous examples are NO_3^- (nitrate), HSO_4^- (bisulphate), HCO_3^- (bicarbonate), NH_2^- (amide), CN^- (cyanide) and OH^- (hydroxide). The extra electrons in these species are quite strongly bound, often by 3 eV or more, and visible radiation is of insufficient energy to detach them. Hence, these very stable negative ions are actually quite difficult to study by the indirect ion-beam spectroscopy methods.

Experimental

A new laser absorption spectroscopy technique, developed recently at Berkeley for studying molecular ions in electrical discharges (Gudeman & Saykally 1984), is very well suited for measuring vibration–rotation spectra of the strongly bound 'classical anions'. In an electrical discharge plasma, Debye shielding of positive ions by the electrons allows a high concentration of charged species to exist; e.g. 0.01 Torr† of positive ions can be obtained in these plasmas. Hence, direct absorption spectroscopy can be used to study ions in plasmas; however, the concentration of neutral molecules in these media is still at least three orders of magnitude larger than that of the ions, and interference from the more abundant neutral absorbers would severely obscure the molecular-ion spectra under ordinary conditions.

The approach developed at Berkeley to separate absorption or emission features of ions from those of the more abundant neutral species has been to take advantage of the Doppler effect that results from the net drift motion of ions in the electric field of a positive column plasma. In a DC positive column plasma, an electric field develops along the axis of the discharge such that positive ions are accelerated toward the cathode and negative ions toward the anode. In the plasma cells used for this work this electric field is about 10 V cm^{-1}. Ions experiencing this field and making many collisions with neutral species before they are neutralised by electrons at the cell walls reach an average drift velocity (V_d), determined by the properties of these collisions, of ca. 500 m s^{-1}, which is about the same magnitude as the average random thermal velocity in the discharge. This net motion results in a Doppler shift in the spectra of ions, given by the usual first-order expression

$$\Delta \nu_{Doppler} = (V_d/C)\nu_{IR}.$$

The magnitude of this shift (ca. 10^{-6} of the infrared transition frequency) is about the same as the linewidth determined by the random thermal motion. By repeatedly reversing the polarity of the plasma we can essentially Doppler shift an absorption line into and out of resonance with a tunable monochromatic infrared laser. Because neutral species do not experience such Doppler effects in their spectra, rapid polarity switching (50000 times per second) and phase-sensitive detection of the laser power allows us to both achieve a very high absorption sensitivity (ca. 10^{-6}) while suppressing the absorptions of the more abundant neutral species by as much as 99.9%. This technique is termed 'velocity-modulation laser spectroscopy'.

Diagrams of the Berkeley velocity-modulation spectroscopy experiments are shown in figures 1 and 2. Tunable infrared laser radiation, either from a Burleigh colour-centre laser (2850–4200 cm^{-1}) or from a Laser Analytics lead-salt diode laser (350–3000 cm^{-1}) is directed through a velocity-modulation sample cell into a suitable detector. With the colour-centre system, the laser beam is split into two components, and one of them impinges on a matched reference

† Torr ≈ 133.3 Pa.

[24]

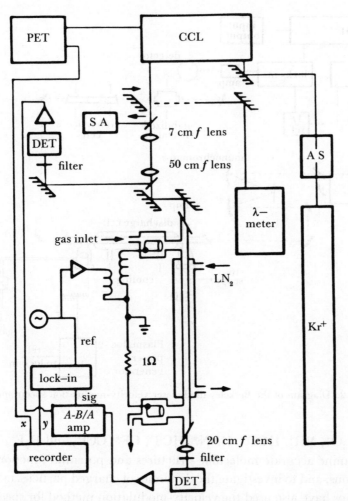

FIGURE 1. Diagram of the Berkeley colour-centre velocity-modulation laser spectrometer.

detector for subtraction of low-frequency amplitude fluctuations. The detector outputs are preamplified, subtracted in a difference amplifier, and demodulated at the discharge frequency to yield the velocity-modulation spectra, which have the usual first-derivative lineshape characteristic of frequency-modulation techniques. The laser frequency in these experiments is measured to high accuracy with either the lines from a calibration gas or with a wavemeter. A microcomputer controls the scanning of the laser and the collection of data. Both infrared laser systems are capable of measuring fractional absorptions as small as one part in a million with an accuracy limited by the Doppler linewidths (*ca.* 10^{-7}).

The velocity-modulation plasma cells are usually 1 cm × 100 cm, can be cooled to 77 K, and operate at pressures in the range 1–50 Torr. The ion density achieved when these plasmas are driven at 500 W is near 10^{14} ions cm^{-3}. Studies of the dynamics of positive ions have shown that rotational temperatures of ions are in equilibrium with the neutral gas kinetic temperature in the centre of the plasma, which is *ca.* 200 K hotter than the cell walls in He or H_2 plasmas because of finite thermal conductivity of the gas mixture.

Over the past several years we have employed the velocity-modulation technique in studies of infrared spectra of a variety of fundamental positive molecular ions (Gudeman & Saykally

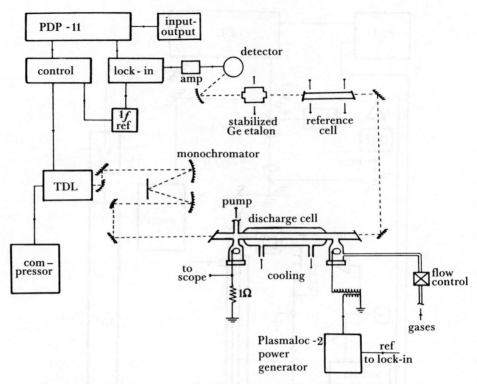

FIGURE 2. Diagram of the Berkeley diode laser velocity-modulation spectrometer.

1984), including H_3O^+, NH_4^+, HCO^+, HNN^+, HCS^+, CF^+, CCl^+, and H_2F^+. The goals of this work were to determine accurate molecular structures and potential functions near the equilibrium configurations, and to investigate the dynamics of charged particles in plasmas. Several other research groups have also used the velocity-modulation method for spectroscopic studies of important molecular cations; most notably, Oka's group at the University of Chicago has studied H_3^+, NH^+, NH_2^+, H_3O^+, H_2O^+, OH^+, NH_4^+, $HCNH^+$, CH_3^+ and possibly $H_3C_2^+$ (protonated acetylene) and Davies's group at Cambridge University has investigated H_3O^+, HCO^+, HCS^+, HCl^+, SH^+ and some other ions. Several other groups have very successfully used direct laser absorption spectroscopy of different types of plasmas (negative glows) for studies of interesting ions, but without this definitive discrimination against neutral molecules. Considering that the first observation of the infrared absorption spectrum of a molecular ion was made by Takeshi Oka as recently as 1980, the fact that more than 25 charged species have now been studied in detail by infrared spectroscopy indicates the tremendous rate of progress in this field (Gudeman & Saykally 1984).

In August 1985, Jeff Owrutsky, a graduate student at Berkeley, discovered the vibration–rotation spectrum of hydroxide in a H_2–O_2–Ar discharge (Owrutsky *et al.* 1985). In an interesting development, the OH^- signals disappeared when the cell was cleaned. It was later determined that the existence of a layer of sputtered metal on the discharge cell wall was necessary to achieve large negative-ion signals (Rosenbaum *et al.* 1986). This same effect was observed for NH_2^- in an NH_3 plasma. The actual role of the sputtered metal is presently unclear, but it may involve a surface-enhanced production of negative ions in these plasmas.

The magnitude of the metal enhancement effect is demonstrated in figure 3. This spectrum

[26]

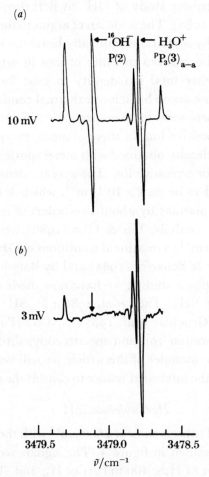

FIGURE 3. Comparison of the P(2) transitions of OH⁻ and the $^{p}P_3(3)_{a-a}$ transition of H_3O^+ observed in a clean discharge cell (b) with those observed in a cell with copper metal sputtered on the cell wall (a). Positive ion signals are not strongly affected by the condition of the cell walls.

also demonstrates two important features of the velocity-modulation technique. In this 1 cm⁻¹ scan, both the incompletely suppressed absorption of a neutral molecule (highest-frequency line) and absorptions from positive ions (H_3O^+) appear, in addition to the OH⁻ lines. These absorptions are easily distinguished; the neutral features become very much larger (greater than 100) if detected at twice the discharge frequency ($2f$) because the plasma turns on and off at this rate, changing the concentrations of most species as it does. The effect cancels out if detection is carried out at the discharge frequency itself ($1f$) except for the residual population changes caused by the inability to make both half-cycles of the ac discharge exactly identical. Also, positive and negative ions exhibit opposite translational motions in the plasma electric field and, as a result, their demodulated first derivative lineshapes have opposite symmetry. Hence it is straightforward to decide whether or not a given absorption feature is caused by a negative or positive ion, or by a neutral species. This capability is an extremely important advantage of the velocity-modulation method, because the plasmas that optimize the concentrations of most negative ions also contain high densities of neutral molecules having strong and extensive infrared absorptions, and the positive ions generated in most plasmas are usually 100 times more abundant than the anions.

[27]

We have carried out an extensive study of OH⁻ at Berkeley, including both the ^{16}O and ^{18}O isotopes (Rosenbaum *et al.* 1986). The addition of argon to the H_2–O_2 plasma enhances the OH⁻ concentration substantially and also rotationally heats the ion. This enhancement of the OH⁻ density is the result of the reduced mobility of ions in argon relative to that found in lighter gases that forces a higher total ion density to exist for a given operating current. Rotational heating arises because argon has a lower thermal conductivity than do lighter gases. Based on these observations, there seem to be no dramatic differences between such dynamical properties of OH⁻ and small positive ions in these plasmas, except for the large enhancement (greater than 10^2) of the OH⁻ density obtained with metal sputtered on the cell walls, an effect that has never been observed for a positive ion. The average density of OH⁻ in H_2–O_2 plasma with argon added is calculated to be *ca.* 2×10^{11} cm⁻¹, which is lower than that observed for simple positive ions in similar plasmas by about two orders of magnitude.

Based on the results of this analysis, Liu & Oka (1986) were nearly able to assign two absorptions observed near 400 cm⁻¹ to rotational transitions of OH⁻. Guided by the theoretical calculations carried out by Lee & Schaefer (1985) and by Botschwina (1986) and by spectroscopic results from condensed phase studies, we have now made extensive studies of a variety of molecular anions, including NH_2^- (Tack *et al.* 1986 *a,b*), SH⁻ (Gruebele *et al.* 1987 *a*), N_3^- (Polak *et al.* 1987 *a*), NCO⁻, (Gruebele *et al.* 1987 *b*), NCS⁻ (Polak *et al.* 1987 *b*) and CCH⁻ (Gruebele *et al.* 1987 *c*) by vibration–rotation spectroscopy, through the use of the velocity-modulation technique. In the remainder of this article we will very briefly describe the results of these studies. We encourage the interested reader to consult the original references for details.

Hydrosulphide (SH⁻)

Representative spectra of SH⁻ observed near 2600 cm⁻¹ with the Berkeley diode laser velocity modulation spectrometer are shown in figure 4. The signals were optimized in a discharge through a mixture of 70 mTorr of H_2S, 400 mTorr of H_2, and 700 mTorr of Ar with the cell

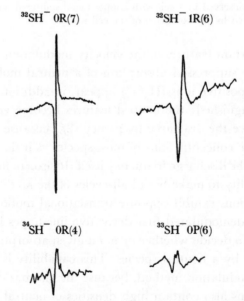

FIGURE 4. Some representative transitions observed for SH⁻. The apparent difference in linewidths is caused by the number of channels per line collected by the computer. The number in front of the rotational assignment refers to the lower vibrational state.

walls cooled to 10 °C. Although no enhancement of the anion density was achieved by coating the cell with metal, signal:noise ratios as high as 200 were nevertheless observed for $^{32}SH^-$, allowing both the $^{33}SH^-$ and $^{34}SH^-$ fundamentals as well as the lowest $^{32}SH^-$ hot band to be measured. All of these spectra were fitted simultaneously to a Dunham-type expression with appropriate isotope scaling relations to obtain the equilibrium molecular constants. From the fitted value of B_e, an equilibrium bond distance of 1.343255 (14) Å† was obtained for SH$^-$, compared to the value 1.340379 (5) Å for the neutral SH radical derived from the infrared laser study of Bernath *et al.* (1983). An estimate of the dissociation energy ($D_e = 3.92$ eV) of the anion was obtained by fitting the spectra to a third-order potential function and calculating the coefficients of a Morse function. This agrees well with the value of 3.95 eV obtained from photoelectron spectroscopy.

Amide (NH_2^-)

An extensive spectrum was observed for NH_2^- in the range 2900–3400 cm^{-1}, as shown in figures 5 and 6, with the Berkeley colour-centre laser system. Although its electron binding energy is relatively low (0.76 eV), the NH_2^- ion was readily generated in discharges containing *ca.* 3 Torr of NH_3; apparently the production of amide ions proceeds by dissociative electron attachment to ammonia, because this process has a large cross section (2×10^{-18} cm^{-2}) that peaks at an energy typical of electrons in such plasmas (5.6 eV). A substantial enhancement of the NH_2^- density was also observed when the cell was coated with copper metal. The observed density of NH_2^- was calculated to be approximately 4×10^{11} cm^{-3}. Of the *ca.* 180 vibration–rotation transitions measured, 117 were assigned and fit simultaneously to a Watson S-reduced hamiltonian for the v_1 (symmetric stretch) and v_3 (asymmetric stretch) bands. Several large perturbations were observed, ostensibly resulting from both Coriolis and Fermi perturbations between the (020) bending state and the two upper stretching states.

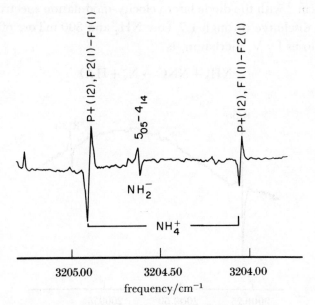

FIGURE 5. Velocity-modulation spectra showing R-branch transition in the v_1 band of NH_2^- and two lines previously assigned to NH_4^+.

† 1 Å = 10^{-10} m = 10^{-1} nm.

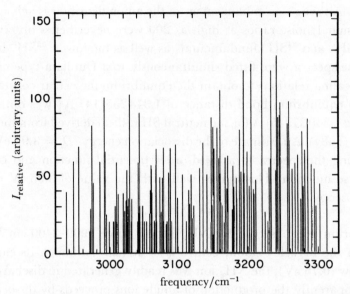

FIGURE 6. Computer plot of all observed lines assigned to NH_2^-.

From the fitted molecular constants, a zero-point geometry was determined; $r_{NH} = 1.0367\,(154)$ Å and $\theta_{HNH} = 102.0\,(3.3)°$. The band origins were: $\nu_1 = 3121.9306\,(61)$ cm^{-1} and $\nu_3 = 3190.291\,(14)$ cm^{-1}. These results are in good agreement with the *ab initio* calculations of Lee & Schaefer (1985) and Botschwina (1986).

Azide (N_3^-)

Figure 7 shows representative spectra of the asymmetric stretching mode of azide anion observed near 2000 cm^{-1} with the diode laser velocity-modulation spectrometer. These spectra were optimized in a discharge through 1.7 Torr NH_3 and 300 mTorr of N_2O. The formation reaction, suggested to us by V. Bierbaum, is

$$NH_2^- + NNO \rightarrow N_3^- + H_2O.$$

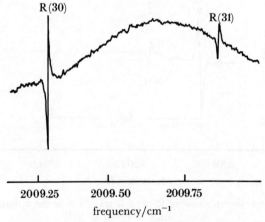

FIGURE 7. Velocity-modulation diode laser spectra of the ν_3 band of N_3^-, showing the 2:1 intensity alternation. The alternation appears larger because of changes in laser power during the scan. The symmetries of the lineshapes identify the transitions as negative-ion absorptions.

[30]

The data presented in figure 7 clearly establish the centrosymmetric nature of N_3^- by virtue of the intensity alternation observed between the adjacent rotational lines; spectra measured in crystalline environments were inconsistent in revealing this property.

The data, ranging from $P(42)$ to $R(62)$, were fitted to a linear molecule effective hamiltonian including quartic terms. The band origin and rotational constants are: $\nu_3 = 1986.4672\,(19)$ cm^{-1}, $B'' = 0.426203\,(57)$ cm^{-1}, and $B' = 0.422572\,(55)$ cm^{-1}, yielding $1.188402\,(82)$ Å as the zero-point bond length. These results indicate a significant increase in the bond length relative to results extracted from X-ray crystallographic studies (1.12–1.16 Å).

Cyanate (NCO⁻)

Figure 8 shows a portion of the ν_3 band (CN stretch) of NCO⁻ observed with the diode laser system in the region near 2100 cm^{-1} in a discharge through NH_3 and CO_2. A variety of hot bands were measured in this work, indicating that the formation reaction suggested by Dr Bierbaum,

$$NH_2^- + CO_2 \rightarrow NCO^- + H_2O,$$

produces high vibrational excitation, particularly in the bending mode.

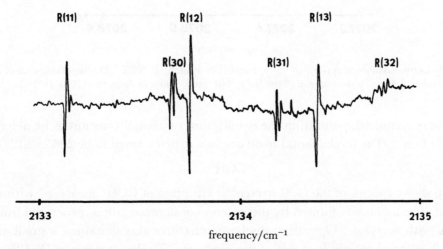

FIGURE 8. Velocity-modulation spectrum of the most intense R-branch region of the ν_3 band (CN stretch) of NCO⁻. The singlets are fundamental lines and the doublets are from the bending hotband.

Over 132 transitions in the CN stretching fundamental and associated bending and stretching hot bands were measured and fitted to a standard linear-molecule hamiltonian, yielding effective parameters for the (00^00), (01^10), (02^00), (10^00), (00^01), (01^01), (02^01), and (10^01) states. Measurement of the rotational constants in all normal modes allowed the determination of the equilibrium rotational constant as $0.385933\,(166)$ cm^{-1} and the ν_3 fundamental was determined to be 2124.307 cm^{-1}.

Thiocyanate (NCS⁻)

The similar formation reaction

$$NH_2^- + CS_2 \rightarrow NCS^- + H_2S$$

was used to produce the NCS⁻ anion in a NH_3–CS_2 plasma. Figure 9 shows representative spectra observed for the CN stretch with the diode laser system near 2078 cm^{-1}. Again, several

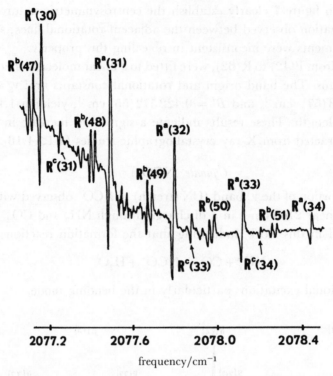

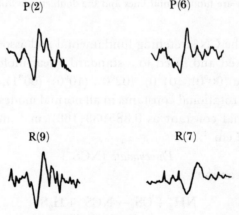

FIGURE 9. Velocity-modulation spectra of the ν_3 band (CN stretch) of NCS$^-$. The superscript label a denotes the fundamental band, b denotes the $(01^10–01^11)$ hot band, and c denotes the $(10^00–10^01)$ hot band.

hot bands were measured, permitting the equilibrium rotational constant to be determined as $0.197\,438\,(61)$ cm^{-1}. The fundamental band origin was determined to be $2065.9302\,(13)$ cm^{-1}.

CCH$^-$

Figure 10 shows spectra of the C–C stretching vibration of CCH$^-$ measured with the diode laser system. The anion was formed by the reaction of fluorine atoms, generated from NF$_3$ in the plasma, with acetylene. Optimized discharge mixtures also contained a small amount of O$_2$, which functioned mainly to stabilize the discharge. Twelve transitions [P(12) to R(11)]

FIGURE 10. Velocity-modulation spectra of the ν_3 band (CC stretch) of CCH$^-$ observed near 1750 cm^{-1}.

[32]

were measured and fitted to a linear-molecule hamiltonian. The C–C stretching band origin was found to be $1758.621\,(3)$ cm^{-1} and the ground-state rotational constant was $1.3814\,(3)$ cm^{-1}.

Botschwina (1986) has calculated the dipole moment of CCH$^-$ to be -3.22 D†. Because of the widespread abundance of neutral CCH in interstellar sources, this makes CCH$^-$ an excellent candidate for an interstellar search. Its detection would certainly lead to important knowledge of the roles of negative molecular ions in the chemistry of interstellar molecular clouds.

CONCLUSIONS

Clearly, the development of high-resolution laser spectroscopy of molecular anions has proceeded at an impressive rate over the last two years. At this stage, the seven anions discussed in this article, as well as the extremely interesting FHF$^-$ anion studied by Kawaguchi & Hirota (1986), have been detected by velocity-modulation spectroscopy. In addition, autodetachment methods have now resulted in the observation of high-resolution spectra for several interesting species (NH$^-$, C$_2$H$_3$O$^-$, CH$_2$CN$^-$, FeO$^-$, PtN$^-$, C$_2^-$).

One of the important reasons for the success of the velocity-modulation method is the impressive knowledge of anion chemistry generated from the flowing afterglow experiments of Bierbaum et al. (1977, 1984) and others. Guided by this work, it appears that a wide variety of negative molecular ions can now be studied in detail by this approach.

An interesting and important result of these laboratory studies of molecular-anion spectra is the generation of a data base for astrophysical searches for negative-ion spectra. Given current estimates of the electron density in cold, dark interstellar clouds, the abundance of several negatively charged molecules such as CCH$^-$, NCO$^-$, OH$^-$ and NH$_2^-$ may be sufficient to enable their detection by millimetre astronomy. In fact, searches are currently continuing by this research group for rotational transitions of CCH$^-$ and NCO$^-$ at several observatories, by using data obtained from the experiments described in this paper. It is hoped that successful detection of several of these anions can be accomplished, and that, as a result, we may begin to explore the currently unknown effects of negatively charged molecules in interstellar chemistry.

This work was supported by the Structure and Thermodynamics Program of the National Science Foundation (grant CHE 8602291) and the NSF Presidential Young Investigator Program. R.J.S. thanks the Berkeley Miller Research Institute for a Fellowship during 1986-1987.

REFERENCES

Al-Za'al, M., Miller, H. C. & Farley, J. W. 1986 *Chem. Phys. Lett.* **131**, 56.

Al-Za'al, M., Miller, H. C. & Farley, J. W. 1987 *Phys. Rev.* A **35**, 1099.

Andersen, T., Lykke, K. R., Neumark, D. M. & Lineberger, W. C. 1987 *J. chem. Phys.* **86**, 1858.

Bernath, P. F., Amano, T. & Wong, M. 1983 *J. molec. Spectrosc.* **98**, 20.

Bierbaum, V. M., DePuy, C. H. & Shapiro, R. H. 1977 *J. Am. Chem. Soc.* **99**, 5800.

Bierbaum, V. M., Grabowski, J. J. & DePuy, C. H. 1984 *J. phys. Chem.* **88**, 1389.

Botschwina, P. 1986 *J. molec. Spectrosc.* **117**, 173.

Gruebele, M., Polak, M. & Saykally, R. J. 1987*a* *J. chem. Phys.* **86**, 1698.

Gruebele, M., Polak, M. & Saykally, R. J. 1987*b* *J. chem. Phys.* **86**, 6631.

Gruebele, M., Polak, M. & Saykally, R. J. 1987*c* *J. chem. Phys.* **87**, 1448.

Gudeman, C. S. & Saykally, R. J. 1984 *A. Rev. phys. Chem.* **35**, 387.

Kawaguchi, K. & Hirota, E. 1986 *J. chem. Phys.* **84**, 2953.

† $1\text{ D} = 3.33 \times 10^{-30}$ C m.

Lee, T. J. & Schaefer, H. F. III 1985 *J. chem. Phys.* **83**, 1784.

Liu, D.-J. & Oka, T. 1986 *J. chem. Phys.* **84**, 2426.

Lykke, K. R., Mead, R. D. & Lineberger, W. C. 1984 *Phys. Rev. Lett.* **52**, 2221.

Mead, R. D., Lykke, K. R., Lineberger, W. C., Marks, J. & Brauman, J. I. 1984 *J. chem. Phys.* **81**, 4883.

Miller, H. C. & Farley, J. W. 1987 *J. chem. Phys.* **86**, 1167.

Neumark, D. M., Lykke, K. R. Andersen, T. & Lineberger, W. C. 1985 *J. chem. Phys.* **83**, 4364.

Owrutsky, J., Rosenbaum, N., Tack, L. & Saykally, R. J. 1985 *J. chem. Phys.* **83**, 5338.

Polak, M., Gruebele, M. & Saykally, R. J. 1987a *J. Am. chem. Soc.* **109**, 2884.

Polak, M., Gruebele, M. & Saykally, R. J. 1987b *J. chem. Phys.* (In the press).

Rosenbaum, N. H., Owrutsky, J. C., Tack, L. M. & Saykally, R. J. 1986 *J. chem. Phys.* **84**, 5308.

Tack, L. M., Rosenbaum, N. H., Owrutsky, J. & Saykally, R. J. 1986a *J. chem. Phys.* **84**, 7056.

Tack, L. M., Rosenbaum, N. H., Owrutsky, J. C. & Saykally, R. J. 1986b *J. chem. Phys.* **85**, 4222.

Phil. Trans. R. Soc. Lond. A **234**, 109–119 (1988)

Printed in Great Britain

Determination of the dipole moments of molecular ions from the rotational Zeeman effect by tunable far-infrared laser spectroscopy

BY K. B. LAUGHLIN, G. A. BLAKE, R. C. COHEN, D. C. HOVDE AND R. J. SAYKALLY

Department of Chemistry, University of California, Berkeley, California 94720, U.S.A.

The details of the first experimental determination of the dipole moment of a molecular ion from the rotational Zeeman effect are presented, along with an assessment of the ultimate accuracy of the technique.

The permanent electric-dipole moment of a molecule is an important property for characterizing its electronic structure and radiative energy-transfer processes. For a large number of neutral molecules, precise measurements (0.01 %) of dipole moments have been carried out exclusively through use of the Stark effect. Because charged molecules are accelerated in an electric field, observation of the Stark effect becomes impractical, and experimental determination of the electric dipole of an ion has not yet been made, although the dipole derivative of CH_2^- has been determined by Okumura *et al.* (1987) from measurements of the radiative lifetimes of its excited vibrational states. As shown by Townes & Schawlow (1955), Townes *et al.* (1975) and Gordy & Cook (1970), measurement of the rotational g-factors (g_r) for two isotopes of a linear molecule allows the determination of both the magnitude and sign of the electric dipole. For asymmetric molecules, the isotopic dependence of the full g-tensor must be evaluated. Such a determination of the dipole moment from the Zeeman effect is intrinsically less accurate because the dipole moment is proportional to the small difference between g_r/B for two isotopes, where B is the rotational constant. There are also systematic errors associated with the zero-point vibrations of the molecule if the dipole is calculated from ground-state g-factors instead of from equilibrium values. However, it provides a direct experimental route to a quantity previously unavailable for molecular ions.

In the present study, we have used a tunable far-infrared (FIR) laser to measure the $J = 1 \leftarrow 0$ transition of ArH^+ and the $J = 2 \leftarrow 1$ transition of ArD^+. The ions were generated in a glow discharge, and magnetic fields up to 4 kG† were used to measure rotational g-factors. The current experiment permitted only a rough determination of the dipole moment of ArH^+. It is shown that the method could be improved, making possible the measurement of molecular ion dipole moments with a precision of about 0.1 D‡.

The magnetic effects described in detail below are most prominent in hydrogenic molecules, for which rotational transitions occur predominantly in the submillimetre and FIR regions of the electromagnetic spectrum. The induced field splittings are quite small, on the order of one part in 10^5, and a spectrometer designed to measure these effects must overcome not only the technological difficulties of spectroscopic work in the FIR, but must also possess the sensitivity, resolution, and stability to analyse such small perturbations of the rotational spectra of reactive

---

† $1\,\mathrm{G} = 10^{-4}\,\mathrm{T}$.
‡ $1\,\mathrm{D} = 3.33 \times 10^{-30}\,\mathrm{C\,m}$.

ions with high precision. The tunable FIR spectrometer we have constructed for this purpose at Berkeley has been used to study a number of reactive intermediates in the FIR, and is similar in design to others described previously (Bicanic *et al.* 1978; Fetterman *et al.* 1978; Farhoomand *et al.* 1985). A schematic outline of the spectrometer is presented in figure 1.

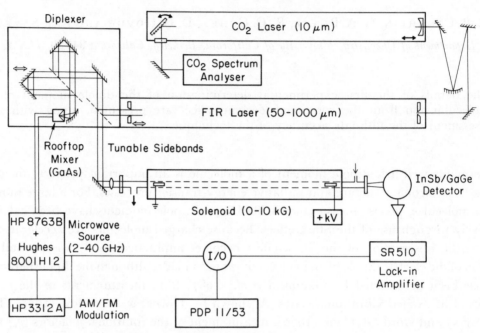

FIGURE 1. Schematic diagram of tunable far-infrared laser spectrometer.

Briefly, by mixing radiation from a fixed-frequency, optically pumped molecular gas laser with that from a tunable microwave oscillator, the resolution and accuracy traditionally associated with conventional microwave spectrometers is upconverted into the FIR were absorption strengths of most molecules are considerably greater. In the initial version of the spectrometer, an FIR laser (1.5 m long, 38 mm bore), optically pumped by a CO_2 laser (1.2 m long, 6.75 mm bore, 40 W output), provided line-tunable radiation of 1–10 mW between 600 and 4000 GHz. A more recent modification of the spectrometer incorporates a commercial 150 W CO_2 laser (Apollo no. 150) and an enlarged FIR cavity (2.3 m, 38 mm bore), which produces line-tunable radiation in the same range, but at greater power levels (50–1000 mW). The microwave radiation that is mixed with the laser is produced by a Hewlett–Packard model 8673B source (2–26.5 GHz) along with a Spacek Ka-2X frequency doubler and a Hughes 8001H12 travelling wave tube amplifier to provide continuous sideband coverage between 2–40 GHz.

A GaAs Schottky-barrier diode (obtained from R. J. Mattauch, University of Virgina) is used to mix the laser and microwave components. The open-structure rooftop mixer in which it is housed is designed to be as broadband as possible at both FIR and microwave frequencies to increase the operational bandwidth of the spectrometer. By sloping the ground plane of the mixer device and contacting the diode with a AuNi long-wire antenna mounted in front of a tunable rooftop reflector (Betz & Zmuidzinas 1984) whose whisper–apex separation may be varied between 100–1000 μm, the FIR optical coupling efficiency is kept nearly constant over a two octave range without creating an efficient retroflector for the laser. The whisker mount mimics a $50\,\Omega$ transmission line except for a short (2 mm) length of wire that gives some flexibility to the antenna. Sufficient microwave radiation can be coupled coaxially up to 23.5 GHz with as little as 10 mW of power. With improved microwave components, coaxial coupling could be extended to much higher frequencies. Frequencies above 26 GHz are coupled onto the diode via a WR-28 waveguide placed in the body of the mixer. Spectra are taken by using either $2f$ sine wave or tone-burst frequency modulation (Pickett 1977) of the microwave source, and are digitally recorded with a PDP-11/53 computer.

The laser radiation is coupled onto the diode after passing through a Martin–Puplett polarization interferometer, which has the advantage of coupling arbitrary laser polarizations onto the diode while simultaneously filtering out the tunable sidebands whose polarization is rotated $90°$ with respect to the laser (Martin & Puplett 1969). A tunable metal mesh Fabry–Perot cavity between the output of the interferometer and the detector increases the filtering of the much more intense laser carrier. Another Fabry–Perot between the laser and corner cube reduces large baselines associated with leaked sideband power being absorbed by sharp resonances with the FIR laser cavity. With the new system, sideband power levels between 100–200 µW have been measured in the 20–30 cm^{-1} region.

A magnetically confined extended negative glow DC discharge was used to produce the ArH$^+$ and ArD$^+$ ions (DeLucia et al. 1983). With this cell, ion densities as high as 5×10^{12} cm^{-3} have been measured for HCO$^+$ at a total pressure of 50 mTorr. Our experiments were performed with a 65 cm long $\times$ 10 cm internal diameter solenoid wound with five layers of 10 gauge magnet wire, shimmed at both ends with two additional layers 5 cm long, and cooled by liquid nitrogen. The maximum usable field was about 4 kG, limited mainly by the cooling efficiency of the magnet and jacket design. Both the cathode and the anode were located entirely between the shim coils, where the measured axial field was homogenous to 1.6 %. The magnet power supply was stabilized by a feedback circuit from a rotating coil field probe (Walker model FFC-4DP), resulting in fields reproducible to 0.01 %. A magnetic field calibration was performed before and after the g-factor measurements by the use of a Hall effect gaussmeter (F. W. Bell, Inc., model 811A). The field calibrations agreed to within 0.15 %. Uncertainties in the magnetic field made a negligible contribution to the error in the dipole moment.

A 60:1 mixture of Ar and H$_2$ optimized the ArH$^+$ absorption at a pressure of 20 mTorr, a discharge current of 12 mA, and with liquid nitrogen cooling of the cell. As found by Bowman et al. (1983), H$_2$ pressures of ca. 1 mTorr or more greatly reduced the absorption. The ArH$^+$ absorption was strongest at about 500 G and decreased somewhat, but not dramatically, at fields up to 4 kG.

Both the ArH$^+$ $J = 1 \leftarrow 0$ and the ArD$^+$ $J = 2 \leftarrow 1$ transitions were measured using the 496 µm (604 GHz) CH$_3$F laser line. For determination of the magnetic splittings, scans were taken over a range of magnetic fields from 800 to 4000 G. Because the magnetic field from the solenoid is perpendicular to the electric field of the FIR radiation, only $\Delta M = \pm1$ transitions are observed. Each spectrum was fitted by a sum of two second derivative lorentzian line shapes, which proved superior to a second derivative gaussian, presumably because of the modulation scheme. Scans for ArH$^+$ and ArD$^+$ are shown in figure 2, and the g-factors from the accumulated data are given in figure 3.

To assess the accuracy of the above data for determination of the rotational g-factors, other magnetic interactions in the molecule must be considered. The magnetic susceptibility anisotropy, $\chi_\parallel - \chi_\perp$, shifts the energy levels of the isoelectronic neutral molecule HCl by only 800 Hz at 4 kG (de Leeuw & Dymanus 1973), and because of the M^2 dependence of the

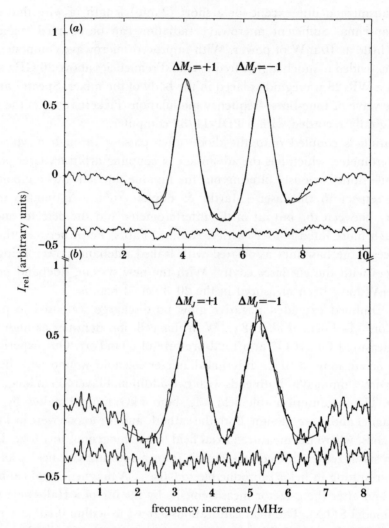

FIGURE 2. Rotational spectra for ArH$^+$ and ArD$^+$ showing the magnetic splitting. The fitted lines are superimposed on the observed spectrum, and the residuals are shown below. (a) ArH$^+$, $J = 1 \leftarrow 0$, frequency from 615.853 15 GHz, B = 2.398 G; (b) ArD$^+$, $J = 2 \leftarrow 1$, frequency from 634.649 21 GHz, B = 3.366 KG.

perturbation it has no effect on the measured splittings. The proton nuclear spin–rotation coupling is 42 kHz for HCl, so that $\Delta M_I \neq 0$ transitions are only very weakly allowed, and the shielding effect for the proton at 4 kG is less than 100 Hz (Kaiser 1970). Therefore the measured splittings should be purely a function of the rotational g-factor, and are given by $\Delta \nu = 2H\beta_1 g_r$, where $H =$ magnetic field and $\beta_1 =$ nuclear magneton (in megahertz per gauss).

The method for calculating the electric dipole from rotational g-factors was first derived by Townes et al. (1955). The following is a short summary of the derivation presented in Gordy & Cook (1970). The rotational g-factor results from the end-over-end rotation of a molecule with imbalanced nuclear and electronic charge. We can express g_r as a sum of nuclear and electronic contributions:

$$g_{xx} = \frac{M_p}{I_x} \sum_k Z_k(y_k^2 + z_k^2) - \frac{2M_p}{mI_x} \sum_{n \neq 0} \frac{|\langle n|L_x|0\rangle|^2}{E_n - E_0} \Bigg\}$$
$$= g_{xx}^{nuc} + g_{xx}^{el}, \tag{1}$$

[38]

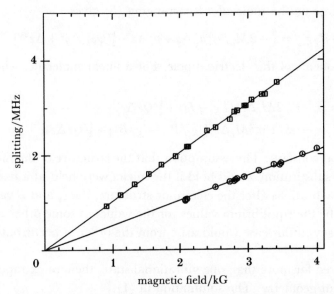

FIGURE 3. Rotational Zeeman-effect splitting against magnetic field for ArH+ and ArH+.
Symbols: □, $g_r(\text{ArH}^+) = 0.675\ 6(17)$; ○, $g_r(\text{ArD}^+) = 0.342\ 5(14)$.

where M_p = mass of the proton, I_x = moment of inertia of the molecule about the x axis, Z_k = charge of nucleus k, y_k and a_k = coordinates of nucleus k, m = mass of the electron, L_x = electronic angular momentum about the x axis, E_n, E_0 = energy of electronically excited state n and ground state 0.

The formula for the magnetic susceptibility

$$\chi_{xx} = \frac{-e^2}{4mc^2} \langle 0 | \sum_i y_i^2 + z_i^2 | 0 \rangle + \frac{e^2}{2m^2c^2} \sum_{n \neq 0} \frac{|\langle n | L_x | 0 \rangle|^2}{E_n - E_0}, \tag{2}$$

where y_i, z_i = coordinates of electron i, can be substituted into (1), giving:

$$I_x g_{xx} = M_p [\sum_k Z_k(y_k^2 + z_k^2) - \sum_i \langle 0|y_i^2 + z_i^2|0 \rangle] - \frac{4mc^2}{e^2} M_p \chi_{xx}. \tag{3}$$

Assuming that an isotopic substitution is made that shifts the origin by $(\Delta x, \Delta y, \Delta z)$ without affecting the structure of the molecule, (3) for the new isotope becomes

$$I_x' g_{xx}' = M_p [\sum_k Z_k(y_k^2 - 2y_k \Delta y + \Delta y^2 + z_k^2 - 2z_k \Delta z + \Delta z^2)$$
$$- \sum_i \langle 0|y_i^2 - 2y_i \Delta y + \Delta y^2 + z_i^2 - 2z_i \Delta z + \Delta z^2|0 \rangle] - \frac{4mc^2}{e^2} M_p \chi_{xx}'. \tag{4}$$

Van Vleck (1932) has shown that the magnetic susceptibility, χ_{xx}, is independent of the origin used. If the principal magnetic axes retain their orientation with respect to the principal inertial axes upon substitution, then $\chi_{xx} = \chi_{xx}'$. Subtracting (3) from (4) gives

$$I_x' g_{xx}' - I_x g_{xx} = -2M_p [\sum_k Z_k(y_k \Delta y + z_k \Delta z) - \sum_i \langle 0|y_i \Delta y + z_i \Delta z|0 \rangle - \tfrac{1}{2}Q(\Delta y^2 + \Delta z^2)], \tag{5}$$

[39]

where Q is the net charge on the molecule. Rearranging (5) gives

$$I'_x g'_{xx} - I_x g_{xx} = (-2M_p/e)[\mu_y \Delta y + \mu_z \Delta z - \tfrac{1}{2}(Qe)(\Delta y^2 + \Delta z^2)], \tag{6}$$

where μ_y, μ_z are components of the electric dipole. For a linear molecule, where $\Delta y = 0$ and $g_{xx} = g_{yy} = g_r$,

$$\mu = -(e/2M_p\Delta z)(I'g' - Ig) + \tfrac{1}{2}(Qe\Delta z)$$
$$= -(eh/16\pi^2 M_p\Delta z)[(g'/B') - (g/B)] + \tfrac{1}{2}(Qe\Delta z), \tag{7}$$

where B is the rotational constant. The assumption that the structure of the molecule remains constant upon isotopic substitution would hold if the nuclei were held at a fixed bond length. Because the zero-point vibrations alter the electronic structure, the g_r and B values used in (6) or (7) must rigorously be the equilibrium values (or the values at some other common point). The limit on the accuracy in this case would arise from the order of perturbation theory used in describing g_r and χ.

If g_r could be measured for more than one vibrational state, then an extrapolation could be made to the equilibrium geometry. The summation in (1),

$$S = \sum_{n \neq 0} \frac{|\langle n|L_x|0\rangle|^2}{E_n - E_0}, \tag{8}$$

is purely a function of the electronic structure, and has the same dependence on bond length for both isotopes if evaluated with the origin at the centroid of the electron distribution. To treat the variation of g_r with the bond length R, Ramsey (1952) assumed the summation S is proportional to $(R/R_e)^l$, where l is an empirical constant. A functional dependence of this type produces the correct behaviour of g_r as $R \to 0$, for $l > 2$. Quinn et al. (1958) were able to correlate their molecular beam measurements of g_r for H_2, HD, and D_2 from this equation, finding $l = 3.7$. The variation of g_r with vibrational state has been measured for SiO, Geo (Davis & Muenter 1974; Honerjager & Tischer 1973) and recently for HF (Bass et al. 1987). SiO and GeO show a very small dependence of g^{el} on v (0.13% and 0.12% respectively for $v = 0$ to $v = 1$). HF, on the other hand, shows a large dependence (11% for $v = 0$ to $v = 1$), two orders of magnitude higher than for SiO and GeO.

Because of the lack of data, the vibrational correction for ArH$^+$ is difficult to estimate. Structurally, ArH$^+$ is more similar to HF than to SiO, and the vibrational dependence of g^{el} should be closer to that for HF. If the summation S in (8) is proportional to R^l, then

$$\langle g^{el}\rangle_{v,J} = g^{el}(R_e)\{1 + (B_e/\omega_e)(v+\tfrac{1}{2})[-3a_1(l-2) + (l-2)(l-3)] + O(B_e/\omega_e)^2)\}. \tag{9}$$

It is important to note that g^{el} in the above equation must be evaluated at the centroid of the electron distribution. However, only for HD (Quinn et al. 1958) and LiH/LiD (Lawrence et al. 1963) are the contributions to g^{el} due to a displacement of the electronic centroid from the centre of mass more than a few percent of the total g^{el}. For most molecules, taking g^{el} to be evaluated at the centre of mass when using (9) to extrapolate back to g^{el} (R_e) introduces negligible error. The g-factors for HF, $v = 0$ and $v = 1$ give $l = 4.74$. With this l value, the linear correction term accounts for 80% of the total correction. If we use this parameter to extrapolate g_r to give $g_r(R_e)$, the dipole moments and corrections due to the extrapolation are given in table 1. To provide an experimental test of the method, calculations based on the literature data for HF, H_2, LiH, CO and GeO are included, and these are compared with the

TABLE 1. ROTATIONAL g-FACTORS AND DIPOLE MOMENTS FOR ArH$^+$ AND HF, H$_2$, LiH, CO AND GeO

(μ is the dipole moment calculated from $g_r(R_e)$, and $\mu = \mu(g_r(R_0)) + \Delta\mu$. μ_0 (Stark) is the experimental value for $v = 0$ obtained from the Stark effect.)

molecule	g_r	$g_r(R_e)$	μ	$\Delta\mu$	μ_0(Stark)	references
ArH$^+$	0.6756 (17)[a]	0.6904[b]	1.59 (40)[a]	0.55	—	—
ArD$^+$	0.3425 (14)[a]	0.3479[b]	1.45 (40)[a]	—	—	—
HF	0.74104 (15)	0.75400[b]	1.93 (44)	0.13	1.826526 (7)	1, 2
DF	0.369 (5)	0.374[b]	—	—	—	3
LiH	−0.654 (7)	—	5.90 (40)	—	5.882 (3)	4, 5
LiD	−0.272 (5)	—	—	—	—	4
^{70}GeO	−0.14179	−0.141415	3.27 (12)	−0.02	3.2824	6, 7, 8
^{74}GeO	−0.14089	−0.140515	—	—	—	6, 7, 8
CO	−0.26910 (20)[c]	—	0.245 (150)	—	0.1098	9, 10
^{14}CO	−0.24664 (20)[c]	—	0.130 (84)	—	—	9
C^{18}O	−0.25622 (20)[c]	—	0.064 (118)	—	—	9
^{13}CO	−0.25704 (20)[c]	—	—	—	—	9
H$_2$	0.88291 (7)	0.89004[d]	—	—	—	11
HD	0.663211 (14)	0.667872[d]	−0.00266 (40)	0.0049	0	12
D$_2$	0.442884 (52)	0.445402[d]	−0.00098 (56)	−0.0063	—	13

[a] Uncertainties are 1σ.

[b] g_r has been extrapolated by using $l = 4.74$.

[c] Uncertainties are relative.

[d] g_c has been extrapolated by using $l = 3.7$.

References: 1, Bass et al. (1987); 2, de Leeuw & Dymanus (1973); 3, Nelson et al. (1960); 4, Lawrence et al. (1963); 5, Wharton et al. (1960); 6, Davis & Muenter (1974); 7, Honerjager & Tischer (1973); 8, Raymonda et al. (1970); 9, Rosenblum et al. (1958); 10, Muenter (1975); 11, Harrick & Ramsey (1954); 12, Quinn et al. (1958); 13 Barnes et al. (1954).

dipole moments obtained from Stark-effect measurements. The LiH and CO g-factors have not been extrapolated to their equilibrium values because the extremely aspherical electron clouds result in negative values of g_r, and these molecules would be expected to have a very different dependence of g_r on bond length than HF or H_2. GeO shows a smaller linear dependence on v, which was used to extrapolate g_r (Davis & Muenter 1974). The result for ArH^+ (corrected for vibrational effects based on the recent HF data), $\mu = 1.59 \pm 0.4$ D (1σ), lies just within the 2σ range of the theoretical value of 2.2 D, calculated by Rosmus (1979) with the coupled electron-pair approximation (CEPA) method.

The dependence of g_r on the bond length will be different for each molecule. From (1), g^{nuc} is completely independent of R, and g^{el} determines the R dependence. At short bond lengths, the electron cloud becomes spherical for $^1\Sigma$ molecules, and $g^{el} \to 0$. For neutral molecules, as the bond stretches to near dissociation, the electron clouds become spherical around their respective nuclei, and $g^{el} \to -g^{nuc}$, so that $g_r \not\to 0$. However, for ions, there will be a residual charge on one atom near the dissociation limit, and $g_r \to 0$. For ArH^+ in particular, dissociating to Ar and a proton, g_r very nearly approaches g^{nuc} at large R. Thus, the dependence of g_r on R near dissociation is radically different for ions than neutrals. Measurement of g-factors in excited vibrational states would yield information on the redistribution of electronic charge as the bond lengthens. ArH^+ is a tightly bound molecule, and near its equilibrium structure the vibrational dependence of g_r is expected to be similar to HF. An experimental confirmation of the vibrational dependence of g_r would remove the doubt involved in the dipole correction.

The vibrational corrections to the dipole moment are extraordinarily large for hydrides. Upon isotopic substitution of deuterium, the reduced mass changes by about a factor of two, resulting in large changes in zero-point vibrational motion. At the opposite extreme, an isotopic substitution of heavier nuclei such as $H^{13}CO^+$ or $HC^{18}O^+$ would have a much smaller change in zero-point vibration, and the corrections mentioned above would probably be negligible. Hydrides such as ArH^+ have very large g-factors, however, making an accurate experimental determination of g_r easier.

The precision of the experimental determination of g_r could be improved both by increasing the magnetic field and by stabilizing the FIR laser. A magnet capable of 10 kG, for example would reduce the error in g_r by a factor of 2.5. The contribution to the error in the dipole moment of ArH^+ would then be 0.15 D (1σ). A magnet capable of producing this field has been constructed, and will be used for future g-factor measurements. FIR laser frequency stability was also a source of error in the determination of g_r. Although the error in the splitting for a single scan was 5 kHz, the r.m.s. deviation of the splitting against field was 25 kHz. If the FIR laser could be stabilized to less than 5 kHz for the duration of a single scan, then the line fit should improve to this level of accuracy. Also a slow linear drift of the FIR laser frequency could be eliminated by taking up and down microwave sweeps, and averaging them. With these improvements, the g-factor uncertainty should decrease by about a factor of twelve. Even more accuracy can be obtained for heavier molecules by observing lower-frequency transitions because of the decreased Doppler linewidth.

For small g-factors that give incomplete splitting of the lines, circular polarization of the radiation can be used to isolate the peaks. This can be accomplished by use of either a quarter wave plate suitable for the absorption frequency, or a polarizing interferometer such as the one used to isolate the sidebands from the laser in the tunable FIR spectrometer. Circular polarization of the radiation also permits determination of the sign of g_r, and thus the sign of the dipole moment. Only one of the $\Delta M = \pm 1$ peaks is observed, and the handedness of the beam can be determined by comparison with a molecule of known g-factor sign, such OCS (Eschbach & Strandberg 1952). In the case of the polarizing interferometer, the beam handedness can be easily decided without a reference molecule.

The Zeeman-effect method for determining electric dipole moments should be applicable to a large number of molecular ions, and the technique has been used to verify the signs of dipole moments for many neutral molecules. In the latter case, a very modest accuracy proves satisfactory because the magnitudes are well known from Stark-effect experiments. If the goal of the experiment is to measure the magnitude as well as the sign, then the accuracy desired for the dipole determines the necessary accuracy for g_r. From (7), the two parameters that determine the uncertainty in μ for a given uncertainty in g_r are Δz, the displacement of the centre of mass upon isotopic substitution, and B, the rotational constant. Larger values of Δz and B reduce the required accuracy in g_r. Table 2 shows the required accuracy in g_r for a 0.1 D uncertainty in μ for various molecular ions and isotopic substitutions. Rosenblum et al. (1958) were able to determine the dipole moment of CO to ± 0.1 D using microwave spectroscopy, and Davis & Muenter (1974) obtained ± 0.12 D accuracy for GeO with molecular-beam techniques. Both of these results agree with the Stark-effect values within experimental error, and the small vibrational correction for GeO (-0.02 D) could result from zero-point changes in the dipole moment. For other molecules, the accuracy is substantially worse, perhaps because of lack of interest. Asymmetric tops present a serious problem at the resolution of conventional rotational spectroscopy, because for a given rotational transition, the different M levels have different splittings. Fitting of lineshapes to sums of unresolved peaks is unlikely to result in the necessary precision for a satisfactory determination of the dipole moment. Unless much higher resolution spectra can be obtained, this technique will be limited to linear and symmetric top molecules.

TABLE 2. REQUIRED PRECISION IN g_r FOR DETERMINATION OF THE ELECTRIC DIPOLE TO ± 0.1 D
FOR VARIOUS MOLECULAR IONS

(Δz is the displacement of the centre of mass for the given isotopic substitution, B is the rotational constant of the heavier ion, and g_r gives the magnetic moment in nuclear magnetons.)

molecule	$\Delta z/\text{Å}$[a]	B/GHz	Δg_r for $\Delta \mu = 0.1$ D	$\Delta g_r H \beta_I$ at 10 kG/kHz
ArH^+/ArD^+	0.030	153	0.00038	5.8
HeH^+/HeD^+	0.11	629	0.0057	87
NeH^+/NeD^+	0.043	281	0.0010	15
$H^{13}CO^+/HC^{18}O^+$	0.053	43	0.00019	2.9
$H^{15}NN^+/HN^{15}N^+$	0.037	45	0.00014	2.1
$H^{13}CS^+/HC^{34}S^+$	0.017	21	0.000030	0.46

[a] 1 Å = 10^{-10} m = 10^{-1} nm.

In summary, we have presented the details of the first experimental determination of the dipole moment of a molecular ion. Our current result for the dipole moment of ArH^+, (1.59 ± 0.4 D), lies just within the 2σ error bar of the ab initio result of Rosmus (1979). It is shown that rotational spectroscopy can be used to achieve the necessary precision for a reasonable determination (± 0.1 D) of the dipole moments of a variety of ions, although extension to asymmetric tops appears unlikely in the near future. Zero-point vibrational effects, which can be substantial for deuterium substitution of hydrides, are likely to be unimportant for other types of molecules.

This work was supported by the National Science Foundation under Grant no. CHE-84-02861. K. B. L. thanks the National Science Foundation for a fellowship. G. A. B. is a Berkeley Miller Research Fellow, 1985–87, and R. J. S. was a Berkeley Miller Research Professor, 1985–86.

REFERENCES

Barnes, R. G., Bray, P. J. & Ramsey, N. F. 1954 *Phys. Rev.* **94** (4), 893.
Bass, S. M., DeLeon, R. L. & Muenter, J. S. 1987 *J. chem. Phys.* **86** (8), 4305.
Betz, A. & Zmuidzinas, J. 1984 In *Proc. Airborne Astr. Symp.* (NASA CP-2353), p. 320.
Bicanic, D. D., Zuidberg, B. F. J. & Dymanus, A. 1978 *App. Phys. Lett.* **32**, 367.
Bowman, W. C., Plummer, G. M., Herbst, E. & DeLucia, F. C. 1983 *J. chem. Phys.* **79**, 2093.
Davis, R. E. & Muenter, J. S. 1974 *J. chem. Phys.* **61** (7), 2940.
DeLucia, F. C., Herbst, E., Plummer, G. M. & Blake, G. A., 1983 *J. chem. Phys.* **78**, 2312.
Eschbach, J. R. & Strandberg, M. W. P. 1952 *Phys. Rev.* **85** (1), 24.
Farhoomand, J., Blake G. A., Frerking, M. A. & Pickett, H. M. 1985 *J. appl. Phys.* **57**, 1763.
Fetterman, H. R., Tannenwald, P. E., Clifton, B. J., Parker, C. D., Fitzgerald, W. D. & Erickson, N. R. 1978 *Appl. Phys. Lett.* **33**, 151.
Gordy, W. & Cook, R. L. 1970 *Microwave molecular spectra.* New York: Interscience Publishers.
Harrick, N. J. & Ramsey, N. F. 1954 *Phys. Rev.* **94**, 893.
Honerjager, R. & Tischer, R. 1973 *Z. Naturf.* A**28**, 1374.
Kaiser, E. W. 1970 *J chem. Phys.* **53**, 1686.
Lawrence, T. R., Anderson, C. H. & Ramsey N. F. 1963 *Phys. Rev.* **130**, 1865.
de Leeuw, F. H. & Dymanus, A. 1973 *J. molec Spectrosc.* **48**, 427.
Martin, D. H. & Puplett, E. 1969 *Infrared Phys.* **10**, 105.
Muenter, J. S. 1975 *J. molec. Spectrosc.* **55**, 490.
Nelson, H. M., Leavitt, J. A., Baker, M. R. & Ramsey, N. F. 1961 *Phys. Rev.* **122**, 856.
Okumura, M., Yeh, L. I., Normand, D., van den Biesen, J. J. H., 1987 Bustamente, S. W., Lee, Y. T., Lee, T. J., Handy, N. C. & Schaefer, H. F. *J. chem. Phys.* **86**, (7), 3807.
Pickett, H. M. 1977 *Rev. scient Instrum.* **48**, 706.
Quinn, W. E., Baker, J. M., LaTourrette, J. T. & Ramsey, N. F. 1958 *Phys. Rev.* **112**, 1929.
Ramsey, N. F. 1952 *Phys. Rev.* **87**, 1075.
Raymonda, J. W., Muenter, J. S. & Klemperer, W. A. 1970 *J. chem. Phys.* **52**, 3458.
Rosenblum, B., Nethercot, A. H. & Townes, C. H. 1958 *Phys. Rev.* **109** (2), 400.
Rosmus, P., 1979 *Theor. chim. Acta* **51**, 359.
Townes, C. H., Dousmanis, G. C., White, R. L. & Schwarz, R. F. 1955 *Disuss. Faraday Soc.* **19**, 56.
Townes, C. H. & Schawlow, A. L. 1975 *Microwave spectroscopy*, New York: Dover Publications.
van Vleck, J. H. 1932 *Theory of electric and magnetic susceptibilities.* Oxford: Clarendon Press.
Wharton, L., Gold, L. P. & Klemperer, W. 1960 *J. chem. Phys.* **33**, 1255.

Discussion

H.-J. FOTH (*University of Kaiserlautern, F.R.G.*). For the example of NCO^-, wherein the bending-mode excitation is much higher than that of the stretching mode, Professor Saykally presented temperatures of 2000 and 700 K, respectively. How big are the uncertainties, and how many vibrational bands have been used for the analysis? What is the rotational temperature?

R. J. SAYKALLY. The vibrational temperatures, as estimated from relative intensities of adjacent fundamental and hot band lines, are probably reliable to about 30 %. A total of eight different vibrational states were sampled. Rotational temperatures, extracted from Boltzmann plots in the respective bands, yield an average temperature of 720 K, reliable to about 30%. Because the neutral translational temperature was determined from Doppler profiles to be near 650 K, we again find support for the idea that the ionic rotational temperature is equilibrated with the neutral kinetic temperature, which is hotter than the cell wall because of the finite thermal conductivity of the gas.

E. HIROTA (*Institute for Molecular Science, Okazaki* 444, *Japan*). I am very much impressed by Professor Saykally's success in determining the dipole moment of ArH^+ ion. Does his method allow him to determine the sign of the g_r factor? The sign of the dipole moment depends on the sign of g_r. Which end of ArH^+ does he think is positive?

R. J. SAYKALLY. The sign of g_r can easily be determined by using a quartz $\frac{1}{4}\lambda$ plate that works at *ca.* 600 GHz. One of the $\Delta M = \pm 1$ peaks would vanish when the FIR beam becomes circularly polarized. By comparison with CO, which is known to have a negative g-factor (Rosenblum *et al.* 1958), the sign of the g-factor of ArH^+ can be verified. We have performed this experiment, yielding $g_r > 0$, and thus a dipole moment of polarity–ArH^+.

General reference

Miller, D. J., DeLeon, R. & Muenter, J. S. 1980 *J. molec. Spectrosc.* **83**, 283.

T. OKA, F.R.S. (*University of Chicago, U.S.A.*). How did Professor Saykally obtain the estimated H_3O^+ ion number density of $1-5 \times 10^{13}$ cm^{-3}? If we assume a current density of, say, 0.2 A cm^{-2}, these very large values correspond to very small electron migration velocity of $10^5-2 \times 10^4$ cm s^{-1}, even if we assume all cations are H_3O^+.

R. J. SAYKALLY. These densities were actually estimated for the HNN^+ and H_3O^+ ions, generated in H_2-N_2 and H_2-O_2 plasmas. Our current densities are approximately 0.5 A cm^{-2}. With theoretical values for the Einstein coefficients and a value of 500 K for the rotational temperature, we do in fact, get densities more near *ca.* 10^{12} cm^{-3}, as Professor Oka suggests. These values are consistent with those obtained by assuming electron-drift velocities in discharges through H_2 or O_2 ($1-4 \times 10^6$ cm s^{-1} or $3-8 \times 10^6$ cm s^{-1} at 2–10 V cm^{-1} $Torr^{-1}$ – v: Engel), which, at our current densities, should result in ion densities on the order of 10^{12} ($0.6-2.5 \times 10^{12}$ cm^{-3} for H_2 and $3-8 \times 10^{11}$ cm^{-3} for O_2). However, uncertainties in the vibrational and translational temperatures of the ions and variations in the gas composition can change these estimates by factors of 1–5.

Phil. Trans. R. Soc. Lond. A **324**, 121–130 (1988)

Printed in Great Britain

Infrared laser spectroscopy of cations

By P. B. Davies

Department of Physical Chemistry, University of Cambridge,
Lensfield Road, Cambridge CB2 1EP, U.K.

The high-resolution infrared diode laser absorption spectra of several cations have been recorded and analysed. Most of the spectra were detected with an AC discharge and the velocity-modulation technique, with frequencies up to 60 kHz. In addition to ions, this method can also detect transient neutral species by recording at twice the modulation frequency (concentration modulation). As an example, spectra of an infrared electronic transition in H_2 are described briefly. The diatomic cationic species discussed are CO^+ ($X^2\Sigma^+$) and SH^+ ($X^3\Sigma^-$), which are compared with the isoelectronic neutral free radicals CN and PH respectively, also in their ground states. Polyatomic cations that have been studied include HCO^+ and H_3O^+. The bending mode of HCO^+ (2_0^1) shows a distinctive Q-branch with many resolved components at Doppler-limited resolution and also some hot-band lines. The HCO^+ spectrum can be combined with microwave measurements in the ground vibrational level to yield predictions for (2_1) rotational transitions for astrophysical searches. The most recent spectra of the ν_2 'umbrella' mode of H_3O^+, the $2^+ \leftarrow 1^-$ band, provide spectroscopic data on vibrational levels approximately 1000 cm^{-1} above the inversion potential maximum.

Introduction

Considerable progress has been made recently in our knowledge of the structure of simple ions. Pioneering work in high resolution spectroscopy by Woods and co-workers in the microwave region (Dixon & Woods 1975) and later by Oka (1980) in the infrared showed it was feasible to generate sufficient ion concentrations in discharges for absorption spectroscopy. These experiments were followed by many other studies of (mainly) cationic species as well as *ab initio* calculations of structural properties such as vibrational frequencies and internuclear distances (Botschwina 1986). Because the experimental techniques are not always wide ranging in their frequency coverage, accurate theoretical predictions, for the infrared region at least, are of considerable value. Alternative experimental data from, for example, matrix isolation or photoelectron spectroscopy are often unavailable. This contribution describes the infrared (IR) diode laser spectra of several free cations (including open-shell species) and compares their features with neutral isoelectronic analogues.

Experimental

The difficulty of distinguishing ions from neutrals in a discharge plasma has been overcome with the introduction of velocity modulation (Gudeman *et al.* 1983) and other techniques. However, if the chemical composition of the discharge is not too complicated and reliable predictions are available it is possible to extract IR spectra of ions without recourse to specially developed modulation techniques. Figure 1 is an example, a spectrum of HCl^+ recorded in

[47]

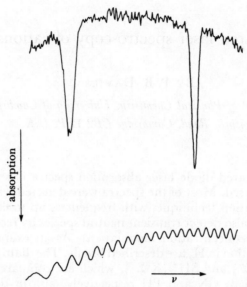

FIGURE 1. Infrared absorption spectrum of the $H^{35}Cl^+$ R(8.5) transition near 2720 cm^{-1} recorded with a 3 m DC discharge as ion source. The apparently unequal intensities of the two components arise from the slightly nonlinear frequency scale.

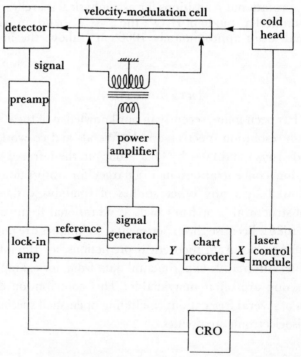

FIGURE 2. Experimental arrangement for velocity (1f) or concentration (2f) modulation detection of transient species generated in an AC discharge.

absorption by rapid-scan diode laser spectroscopy under computer control (Davies *et al.* 1983). This is a particularly favourable case, however, because very accurate electronic spectroscopy provides unambiguous assignments. This method of detection has now been superseded by velocity modulation. Measurement of noise at the detector with a spectrum analyser (Rothwell 1985) shows clearly that the frequency of the AC discharge used in velocity-modulation experiments should be greater than 10 kHz. Figure 2 shows schematically the arrangement for velocity-modulation detection. The modulation wave form used is sinusoidal rather than square wave, and although this yields a lower 'duty cycle' for the discharge there is usually much less distortion of the waveform by the transformer.

An additional benefit of using an AC discharge is that tuning the phase sensitive detector to $2f$, twice the modulation frequency, a concentration modulation effect is observed and species with lifetimes shorter than $1/f$ (neutral or charged) can be detected with 'zeroth-order' absorption lineshapes. Figure 3 is a spectrum of molecular hydrogen observed in this way. The

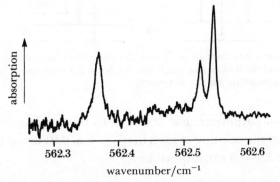

FIGURE 3. Absorption spectrum of excited H_2 generated in an AC discharge and detected by concentration modulation (at $2f$). The transition is the $Q(2)$ component $\Delta v = 0$, $v = 2$ band of the a ← c system. This is a transition in para-H_2 and shows fine structure but no hyperfine structure.

spectrum is part of the infrared a–c electronic transition in H_2 which can be observed in absorption or emission in a hydrogen plasma (Davies *et al.* 1987). Finally, the potential of using an AC discharge to investigate dynamical in addition to purely spectroscopic behaviour has yet to be fully realized. Preliminary work on infrared transitions in Rydberg states of the O atom has shown that the decay of transient species can be followed in real time as the discharge switches on and off (P. R. Brown and P. A. Hamilton, unpublished results).

RESULTS

The four diatomic molecules described here are all free radicals and have quenched orbital angular momentum. The structure of their rotational and fine-structure levels can be classified as Hund's case (b). One pair have $^2\Sigma$ ground state and the other $^3\Sigma$. Figure 4 is a schematic energy level diagram for both cases. The infrared spectra consist of P-branches and R-branches only and for each rotational component $\Delta J = \Delta N$ transitions are usually the most intense except near the band origin.

CN, CO$^+$ (X $^2\Sigma^+$)

For the R- and P-branches of the fundamental band the doubling of each line in the spectrum ($\Delta J = \Delta N = \pm 1$) is determined by the interaction of the unpaired electron with the rotation of the molecule, $\gamma_v N \cdot S$, and is given by

$$N(\gamma_1 - \gamma_0) + \tfrac{1}{2}(3\gamma_1 - \gamma_0), \quad \text{R-branch,}$$
$$N(\gamma_1 - \gamma_0) - \tfrac{1}{2}(\gamma_1 + \gamma_0), \quad \text{P-branch,}$$

[49]

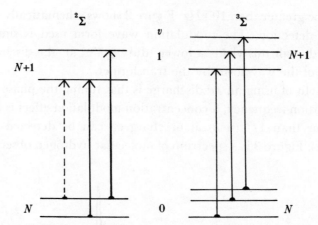

FIGURE 4. Fine structure of diatomic free radicals in $^2\Sigma$ and $^3\Sigma$ states. The transitions shown by solid vertical lines are $\Delta J = \Delta N$ components and the broken line represents the $\Delta J \neq \Delta N$ component (in the $^2\Sigma$ case). The energy levels and transitions are appropriate for the R-branch.

where γ is the spin-rotation parameter. For $2 < N < 10$ these expressions are approximately independent of rotation and the splitting of each line is $\bar{\gamma}$ ($\bar{\gamma} \approx 0.01$ and 0.007 cm^{-1} for CO$^+$ and CN respectively). For the $\Delta J = 0$ satellite we have

$$(\gamma_0 N + \tfrac{1}{2}\gamma_0) \text{ lower,} \quad \text{R-branch,}$$
$$(\gamma_0 N + \tfrac{1}{2}\gamma_0) \text{ higher,} \quad \text{P-branch,}$$

where 'lower' and 'higher' refers to the position relative to the main $\Delta J = \Delta N$ pair. The satellite line gains in intensity at lower N and for R(1) and P(2) is sufficiently close to the $\Delta J = \Delta N$ pair that triplets rather than doublets appear. At R(0) and P(1) only one $\Delta J = \Delta N$ transition is possible but for these rotational components doublet fine structure is still observed because of the satellite line.

Figure 5 shows the R(0) line of CN and, inset, the predicted relative intensities in the fine

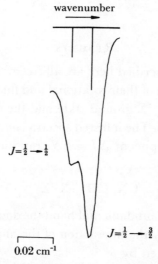

FIGURE 5. Diode laser spectrum of the R(0) transition of the fundamental band of CN(X$^3\Sigma^+$) with resolved fine-structure components. Calculated positions and their relative intensity are shown at the top of the figure. The spectrum was recorded by using rapid-scan spectroscopy under computer control (Davies *et al.* 1983).

[50]

structure (Davies & Hamilton 1982). Many other transitions were observed with doublet splittings given by the formula above. Figure 6 shows the R(1) component of CO⁺ with the predicted triplet fine structure (Davies & Rothwell 1985). (The different lineshapes for CN and CO⁺ reflect the different modulation techniques used.)

$$\text{PH, SH}^+ \ (X \ ^3\Sigma^-)$$

As shown in figure 4, the main fine structure expected in the fundamental band of these molecules is a triplet splitting. Figure 7 shows the P(7) component of PH. The radical was formed in a DC discharge of molecular hydrogen over small amounts of elemental phosphorus (Anacona *et al.* 1984). The size of the spin–spin parameter, λ, in the ground state of PH results

a b c

wavenumber

FIGURE 6. The R(1) component of the fundamental band of CO⁺ $(X^2\Sigma^+)$ recorded with velocity-modulation detection. Peaks b and c are the $\Delta J = +1$ components, peak a is the single $\Delta J = 0$ transition. The calculated relative intensities are a:b:c, 1:5:9.

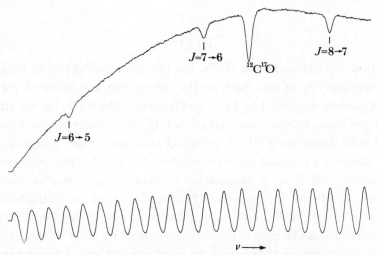

$J=7\rightarrow6$ $^{12}C^{17}O$ $J=8\rightarrow7$

$J=6\rightarrow5$

$\nu \longrightarrow$

FIGURE 7. Diode laser absorption spectrum of the P(7) transition in PH $(X^3\Sigma^-)$ showing the three fine-structure components. The calibration line from $^{12}C^{17}O$ in natural abundance is provided by CO at low pressure in a 15 cm long reference cell.

in a triplet fine structure approximately the same in magnitude as the wavenumber coverage of a single mode of this particular diode. In SH^+, λ is larger (SH^+, $\lambda \approx 5.78$ cm^{-1}; PH, $\lambda \approx 2.21$ cm^{-1}) but triplet structure is still visible, and figure 8 shows the fine structure of one of many components recorded in the fundamental band of SH^+ (Brown *et al.* 1986). Also shown in figure 8 is an additional line due to another cationic species. Possible candidates are electronically excited SH^+ (i.e. $^1\Delta$ or $^1\Sigma$), H_2S^+ or H_3S^+. On the right of the spectrum is another first-derivative line due to an ion but this peak is of opposite phase and arises from SH^- ($X\,^1\Sigma^+$).

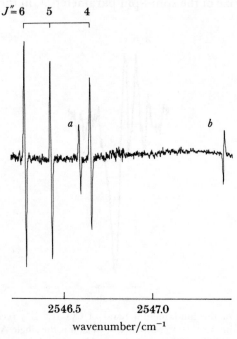

FIGURE 8. Velocity-modulation spectrum around the R(5) region of SH^+ ($X^3\Sigma^-$) showing the expected triplet fine structure. Line *a* arises from another cationic species and line *b* is a vibration–rotation transition in SH^- ($X^1\Sigma^+$).

HCO$^+$

This linear cation, isoelectronic with HCN, has played a central role in the development of high-resolution spectroscopy of ions both in the microwave and infrared regions. All three fundaments have now been studied by laser spectroscopy. However, for the stretching modes individual rotational lines are well separated ($\bar{B} \approx 1.48$ cm^{-1}) and only one transition at a time can be recorded with diode lasers (there are no Q-branches). The Q-branch of the bending mode, however, presents a compact but well-resolved feature that can be recorded with good signal to noise ratios with velocity modulation detection (figure 9). In addition to intense fundamental lines, additional features from hot-band transitions, indicated by *, are also present in figure 9. As well as the purely spectroscopic study of this ion (Davies & Rothwell 1984), the IR laser results can be used to provide predictions for rotational transitions in excited vibrational states for astrophysical purposes. By combining three microwave measurements in the ground state of the ion with six diode laser measurements it is possible to arrive at predictions for two rotational transitions in the 2_1 level. The relevant measurements are given in table 1.

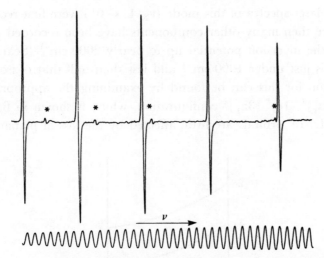

FIGURE 9. Part of the Q-branch of the ν_2 fundamental of HCO^+ near 830 cm^{-1}.

At the time these IR measurements were made, the most accurate calibration procedure available was to use standard gas-calibration lines and germanium etalons as interpolation devices. This limited the accuracy to ± 0.001 cm^{-1}, which was insufficient to yield an unambiguous prediction for radioastronomical purposes (L. Ziurys, personal communication). However, remeasurement of the IR transitions with a confocal Fabry–Perot etalon should improve the accuracy of the IR laser measurements by up to a factor of ten.

H_3O^+

This cation is probably the best example where experiment and theory complement each other. The ν_2 umbrella mode of the molecule has been of particular interest to spectroscopists

TABLE 1. PREDICTED ROTATIONAL TRANSITION FREQUENCIES IN VIBRATIONALLY EXCITED HCO^+

Diode laser results for the 2_0^1 band

line	measured wavenumber[a]	number of measurements	$1\,\sigma$	calculated wavenumber[b]	difference
P(4)	816.3224	3	0.0003	816.3222	+0.0002
P(3)	819.3015	7	0.0007	819.3006	+0.0009
Q(2)	828.2694	6	0.0010	828.2685	+0.0006
Q(3)	828.3066	6	0.0011	828.3056	+0.0010
R(1)	834.1757	5	0.0018	834.1761	−0.0004
R(2)	837.1466	7	0.0018	837.1463	+0.0003

Rotational transitions from ground-state microwave spectroscopy[c]

$J''+1 \leftarrow J''$		wavenumber/cm^{-1}
4	3	11.899375 (2)
3	2	8.924 7615 (3)
2	1	5.949952 (2)

Predicted line positions in the 2^1 level

$3^+ \leftrightarrow 2^-$ 8.9620 ± 0.001 cm^{-1}
$3^- \leftrightarrow 2^+$ 8.9203 ± 0.001 cm^{-1}

[a] P. A. Martin (unpublished results).
[b] Davies & Rothwell (1984).
[c] Sastry et al. (1981).

and theoreticians. IR laser spectra of this mode (ν_2, $1^- \leftarrow 0^+$) were first recorded by Haese & Oka (1984) and since then many other components have been recorded providing a rather complete picture of the inversion potential up to nearly 2000 cm^{-1} (Sears *et al.* 1985). The barrier to planarity is just under 1000 cm^{-1} and less than half that of isoelectronic NH_3. A qualitative explanation for this can be found by examining the appropriate Walsh orbital diagram for the...$(2a_1)^2$ $(1e)^4$ $(3a_1)^2$ configuration, which is shown in figure 10. The $2a_1$–$2a_1'$ and degenerate $1e$–$1e'$ orbitals are little affected by the loss of planarity and are gently

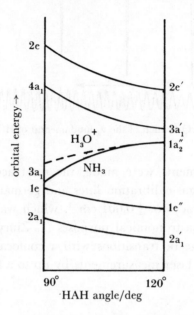

FIGURE 10. Walsh molecular orbitals for AH_3 species.

repulsive as the molecule becomes non-planar. The $3a_1$–$1a_2''$ orbital, comprising an out-of-plane 2p orbital on the heavy atom and H 1s orbitals, is strongly bonding in a non-planar configuration. In H_3O^+ (in comparison with NH_3), the positive charge located on the O-atom contracts the 2p orbital and raises the energy of the $3a_1$–$1a_2''$ molecular orbital relative to its position in NH_3, as indicated in figure 10. Numerous quantitative calculations of the inversion potential have been made (Botschwina 1986) and provide vibrational frequencies and bond lengths essential for experimental spectroscopic searches and interpretation of the very extensive infrared spectrum of H_3O^+.

The results reported here were obtained in collaboration with many colleagues whose names appear in the referenced work. I express my warmest thanks to them for their participation in this work and for many stimulating discussions.

REFERENCES

Anacona, J. R., Davies, P. B. & Hamilton, P. A. 1984 *Chem. Phys. Lett.* **104**, 269–271.
Botschwina, P. 1986 *J. chem. Phys.* **84**, 6523–6524.
Brown, P. R., Davies, P. B. & Johnson, S. A. 1986 *Chem. Phys. Lett.* **132**, 582–584.
Davies, P. B. & Hamilton, P. A. 1982 *J. chem. Phys.* **76**, 2127–2128.

Davies, P. B., Hamilton, P. A., Lewis-Bevan, W. & Okumura, M. 1983 *J. Phys. E.* **16**, 289–294.
Davies, P. B. & Rothwell, W. J. 1984 *J. chem. Phys.* **81**, 5239–5240.
Davies, P. B. & Rothwell, W. J. 1985 *J. chem. Phys.* **83**, 5450–5452.
Davies, P. B., Guest, M. A. & Johnson, S. A. 1987 (In preparation).
Dixon, T. A. & Woods, R. C. 1975 *Phys. Rev. Lett.* **34**, 61–63.
Gudeman, C. S., Begemann, M. H., Pfaff, J. & Saykally, R. J. 1983 *Phys. Rev. Lett.* **50**, 727–731.
Haese, N. N. & Oka, T. 1984 *J. chem. Phys.* **80**, 572–573.
Oka, T. 1980 *Phys. Rev. Lett.* **45**, 531–534.
Rothwell, W. J. 1985 Ph.D. thesis, University of Cambridge.
Sastry, K. V. L. N., Herbst, E. & DeLucia, F. C. 1981 *J. chem. Phys.* **75**, 4169–4170.
Sears, T. J., Bunker, P. R., Davies, P. B., Johnson, S. A. & Spirko, V. 1985 *J. chem. Phys.* **83**, 2676–2685.

Discussion

R. N. DIXON, F.R.S. (*School of Chemistry, University of Bristol, U.K.*). Dr Davies has commented that the inversion barrier of H_3O^+ is less than half of the isoelectronic NH_3 molecule; and it has been proposed that similar comparisons may be made between pairs of ions and molecules of the general type $AH_2^{(+)}$. It has been suggested that this might be attributed to a variation in the angular dependence of the orbital energy for the highest occupied orbital.

However, it must be pointed out that the excitation energy for the $\tilde{A}^2A_1$–$\tilde{X}^2B_1$ transition of H_2O^+ is little different from that of NH_2; this would not be the case on the above postulate. I suggest that this is a simple electrostatic effect. The substitution of N by O^+ will tend to polarize the whole electronic structure by withdrawing electrons from around the H nuclei. These nuclei are therefore less shielded, so that H–H nuclear repulsion will tend to open up the HAH angles, decreasing the inversion barriers between all such isoelectronic pairs in equivalent electronic states. On this basis it seems probable that CH_3^- would be more strongly pyramidal than NH_3, with a larger barrier to inversion in its ground electronic state: these shifts should be smaller than between H_3O^+ and NH_3 because of an increasing AH bond length along the series H_3O^+, NH_3 and CH_3^-.

P. B. DAVIES. The relative contributions of orbital energy and electrostatic effects in determining the inversion barrier in $H_3A^{(+)}$ molecules cannot be readily separated. The energy of the highest-occupied (Walsh) orbital, $3a_1 - 1a_2''$, is strongly angular dependent and is likely to play a major part in determining geometry. The contraction of the heavy atom p-orbital contributing to this molecular orbital as N (in NH_3) is replaced by O^+ (in H_3O^+) makes it less bonding in the non-planar configuration. This effect is in the same direction as the electrostatic effect suggested by Professor Dixon.

R. J. SAYKALLY (*University of California, Berkeley, U.S.A.*). Regarding Dr Davies's analysis of the $\nu_2 = 2$ levels of H_3O^+, has he included Coriolis mixing in his hamiltonian? We have recently completed an analysis of the ν_4 band, and find that the upper ν_4 levels are strongly Coriolis mixed with nearly ν_2 states, resulting in large changes in the rotational constants.

P. B. DAVIES. In our analysis of the $\nu_2 2^+ \leftarrow 1^-$ band, four of the lines measured were poorly fitted and excluded from the final least squares fit. We gave the most likely explanation as Coriolis interaction with the $\nu_4 = 1$, recognizing that this effect could only be fully accounted for when ν_4 band data were available. It is also worth noting that the signs of the centrifugal distortion parameters for the 2^+ level may also affect this. Nevertheless the published parameters are a reasonable representation of the 2^+ level for the lines we have measured (r.m.s. error = 0.006 cm^{-1}).

Reference

Davies, P. B., Johnson, S. A., Hamilton, P. A. & Sears, T. J. 1986 *Chem. Phys.* **108**, 335.

R. J. SAYKALLY. Dr Davies described his measurements on the bending hot band of HCO$^+$, designed to provide precise constants for rotational transitions in the lowest excited bending mode to aid radio astronomy searches. My group has recently completed measurements of several pure rotational transitions in the $v_2 = 1$ state of HCO$^+$ by tunable far-infrared laser spectroscopy. We can now predict all astrophysically important transitions in the first excited bending state with about 100 kHz uncertainties, clearly sufficient precision to identify radio-astronomical spectra in this state.

Reference

Blake, G., Laughlin, K., Cohen, R., Busarow, K. & Saykally, R. J. 1987 *Astrophys. J.* (In the press.)

P. B. DAVIES. The recent introduction of confocal etalons for calibrating diode laser spectra will lead to a considerable improvement in measurement accuracy over methods using germanium etalons (accuracy of order 30 MHz). With a confocal Fabry–Perot etalon it has been shown that diode laser frequency stability of 100 kHz can be achieved. However, linewidths in the far-infrared region are lower than in the mid-infrared, which intrinsically contributes to better measurement precision.

Reference

Reich, M., Schieder, R., Clar, H. J. & Winnewisser, G. 1986 *Appl. Opt.* **25**, 130.

Phil. Trans. R. Soc. Lond. A **324**, 131–139 (1988)

Printed in Great Britain

Infrared diode laser and microwave spectroscopy of molecular ions

By E. Hirota

Institute for Molecular Science, Myodaiji, Okazaki 444, Japan

Some of the recent results on FHF^-, HBF^+, H_2Cl^+, H_3S^+, HCO^+, and DCO^+ are reviewed. The last two molecules are studied to detect the effect of the discharge electric field on l-type doublets in the $v_2 = 1$ and $v_2 = 2$ states. The Fermi interaction between v_1 and $4v_2^0$ in DCO^+ is also discussed. These experimental results are followed by introduction of a simple procedure, based upon the Hellmann–Feynman theorem, that allows us to estimate the electron distributions in ions in comparison with those in related neutrals. Three isoelectronic series (1) NH^-, OH and FH^+, (2) HBF^+, HBO, HCO^+, HCN, HNN^+, HNC and HOC^+, and (3) NH_2^-, OH_2 and FH_2^+ are discussed by this method.

Introduction

Remarkable progress has recently been made on infrared (IR) and microwave (MW) spectroscopy of molecular ions. A number of new species such as protonated ions have been detected and their structures have been explored in detail. Still many interesting ions remain to be observed; examples include the protonated methane and the protonated formaldehyde. We have been trying to detect new ionic species by using either infrared diode laser spectroscopy or microwave spectroscopy, and have recently added a few examples to the list, as described in the present paper. There are a number of ionic species that have already been reported to exist, but still require further studies to unveil unique features of their properties. We have been making effort to study such ions by high-resolution spectroscopy. It may be worth mentioning that no established method has been reported to determine the dipole moment of the ion.

As more data are accumulated on molecular ions, we naturally wonder what really characterizes the charged species. The net charge of the ion plays important roles when it interacts with the environment. The ion is accelerated by the electric field, for example. It has an exceptionally large cross section, i.e. Langevin cross section for reactions with neutral molecules, the reactions known as the ion–molecule reactions, which proceed with almost no activation energy. The next question is how the net charge of the ions affects their properties when they are isolated in vacuum. A clue to answer this question may be provided by comparing molecular parameters of ions with those of isoelectronic neutrals. Recent extensive studies with infrared and microwave spectroscopy will make it possible to perform such comparisons, and so may open a new field in chemistry.

The first part of the present paper summarizes some of the recent results of my group, paying attention to unique features of each ion. The second part describes the development of a new method for estimating charge distribution in molecules and its application to three isoelectronic series of ions and neutrals.

RECENT INFRARED AND MICROWAVE SPECTROSCOPIC STUDIES OF MOLECULAR
IONS

The FHF⁻ ion

Since we published the infrared spectrum of the ν_3 band of this ion (Kawaguchi & Hirota 1986 b), no substantial progress has been made on this ion. One of the most important problems that still remains to be solved for this ion is that *ab initio* calculations cannot reproduce the observed band origin (K. Yamashita, personal communication); the calculated values are lower than the observed value by 300–500 cm⁻¹. We have attempted to detect the FHCl⁻ ion, but have not been successful.

The HBF⁺ ion

Based on our result on the ν_3 band (Kawaguchi & Hirota 1986a), the microwave spectrum of this ion has been observed by two groups (Cazzoli *et al.* 1986, 1987; Saito *et al.* 1987). The measurements were limited to the ground state, and the so-called r_s (substitution) structure was calculated. Vibrational satellites could, however, be observed without too much difficulty. They will allow us to derive not only the equilibrium structure, but also other interesting information on the ion, as demonstrated for HCO⁺ and DCO⁺ (see below). The ν_1 band has recently been observed by K. Kawaguchi & T. Amano (personal communication).

The H₂Cl⁺ ion

We have observed and analysed the ν_2 band (Kawaguchi & Hirota 1986c). The molecular constants of the ion are very similar to those of an isoelectronic molecule H_2S. Kawaguchi & Amano (1987b) have recently extended the measurement to the ν_1, ν_3, and $2\nu_2$ bands; these modes were found to be coupled.

The H₃S⁺ ion

Amano *et al.* (1987) have measured the ν_2 band by infrared diode laser spectroscopy combined with magnetic-field modulation. The ion was generated by a dry-ice-cooled hollow cathode discharge in a mixture of H_2S and H_2 with the partial pressures of 40 mTorr and 1 Torr,† respectively. Interest lay in whether the ν_2 band showed the effect of inversion, but the observed band did not exhibit any indication of splitting. The observed spectrum was analysed in a straightforward way, yielding rotational and centrifugal distortion constants. An anomalous feature of the derived parameters is that D_J, D_{JK}, and D_K change sign on going from the ground state to the ν_2 state. This was explained by the Coriolis interaction with ν_4, which was estimated to be about 160 cm⁻¹ higher in frequency than ν_2. The ν_4 band has not been observed.

The HCO⁺/DCO⁺ ions in excited vibrational states

Kawaguchi *et al.* (1986) studied the ν_1 band of DCO⁺ and found the α_1 constant to be about 100 MHz smaller than that estimated from those of related molecules. They ascribed this observation to the Fermi interaction between ν_1 and $4\nu_2^0$. However, no observation had been reported on the ν_2 state of DCO⁺, preventing a quantitative analysis being done.

Hirota & Endo (1987) observed vibrational satellites of DCO⁺ with a millimeter-wave

† 1 Torr ≈ 133.3 Pa.

spectrometer; the $J = 3 \leftarrow 2$, $4 \leftarrow 3$, and $5 \leftarrow 4$ transitions in the ν_2, $2\nu_2$, ν_1, and ν_3 states were recorded. The observed spectra were analysed by the procedure that De Lucia & Helminger (1977) developed to analyse the vibrational satellites of HCN. The $\nu_2 = 1$ spectra yielded the rotational constant B_{0110}, the centrifugal distortion constant D_{0110}, and the l-type doubling constant q_2, whereas the B and D constants in the 02^00 state and q_2^2/δ were derived from the $\nu_2 = 2$, $l = 0$ spectra, where δ denotes the vibrational energy difference between the $l = 2(\Delta)$ and $l = 0(\Sigma)$ sublevels in the $\nu_2 = 2$ manifold (the suffixes of B and D represent ν_1, ν_2, l, ν_3).

The Fermi interaction between ν_1 and $4\nu_2^0$ was analysed in the following way. The two vibrational states were designated as a and b for the sake of simplicity, and to the unperturbed parameters of the two states were attached a superscript 0. Because the ν_1 satellites were well fitted to a $J(J+1)$ expansion, the unperturbed vibrational energy difference $\Delta E = E_a^0 - E_b^0$ and the interaction term W were assumed to be much larger than the rotational energy $B^0 J(J+1)$. Then the perturbed rotational and centrifugal distortion constants were given by

$$B_a = B_a^0 - (1 - \Delta E/2\Delta)(\tfrac{1}{2}\Delta B), \tag{1}$$

$$D_a = D_a^0 - (1 - \Delta E/2\Delta)(\tfrac{1}{2}\Delta D) - (W^2/8\Delta^3)\,\Delta B^2, \tag{2}$$

where
$$\Delta B = B_a^0 - B_b^0, \quad \Delta D = D_a^0 - D_b^0,$$

and
$$\Delta = [(\tfrac{1}{2}\Delta E)^2 + W^2]^{\tfrac{1}{2}}. \tag{3}$$

The B_a^0 constant, i.e. the unperturbed B constant in the ν_1 state, B_{1000}, was approximated by B_{000} minus the α_1 constant of DCN (325 MHz) (Winnewisser et al. 1971). This α_1 value is very close to the one that Kawaguchi et al. (1986) presumed to be the unperturbed α_1 constant of DCO^+ by reference to the data on related molecules. The B_b^0 constant was given by

$$B_b^0 = B_{0400} = B_{0000} + 2(B_{0200} - B_{0000}) + 8\gamma_{22}. \tag{4}$$

The ground-state B constant, i.e. the first term of (4), was fixed at the value obtained by averaging those reported by Bogey et al. (1981) and Sastry et al. (1981) according to the uncertainties of their measurements. Other ground-state parameters were obtained in the same way. The present result was substituted in the second term of (4), whereas the third term was estimated by using the γ_{22} constant of DCN (1.094 MHz) (Winnewisser et al. 1971). The difference in B, ΔB, was thus obtained to be -729 MHz. The observed B_a constant thus led to $\Delta E/2\Delta = 0.732$. The observed D_a constant was analysed in a similar way, where the change of the D constant on ν_1 excitation, β_1, was estimated to be -0.00051 MHz, which was the value reported for DCN (Winnewisser et al. 1971), and D_b^0 was calculated by the following equation:

$$D_b^0 = D_{0000} + 4(D_{0110} - D_{0000}) + 3q_2^2/\delta. \tag{5}$$

The difference in D, ΔD, was thus calculated to be -0.32535 MHz, and the observed D_a constant yielded $W^2/\Delta^3 = 0.731\,10^{-6}$ MHz^{-1}, which, when combined with $\Delta E/2\Delta = 0.732$ as obtained above, led to $\Delta = 21.1$, $\tfrac{1}{2}\Delta E = 15.5$, and $W = 14.4$ cm^{-1}. The coefficients α and β of the unperturbed wavefunctions are given by

$$\alpha^2 = [\Delta + \tfrac{1}{2}\Delta E]/2\Delta \quad \text{and} \quad \beta^2 = [\Delta - \tfrac{1}{2}\Delta E]/2\Delta. \tag{6}$$

The above result $\Delta E/2\Delta = 0.732$ corresponds to $\beta^2 = 0.134$ and to the unperturbed vibrational levels of $E_a^0(\nu_1) = 2578.9$ cm^{-1} and $E_b^0(4\nu_2) = 2547.9$ cm^{-1}. The ν_2 frequency was estimated to be 660.2 cm^{-1} from the observed l-type doubling constant. Four times this value is 2641 cm^{-1}, which is higher than $E^0(4\nu_2)$ by about 100 cm^{-1}; this is of reasonable magnitude as an anharmonicity correction.

It is interesting to note that we observed only the $l = 0(\Sigma)$ component in the $2\nu_2$ spectrum, for both HCO$^+$ and DCO$^+$. This fact is explained by inhomogeneity of the discharge electric field in the hollow cathode, which broadens the $l = 2(\Delta)$ components. The effect of such Stark effect, although to much less extent, was also observed for the l-type doublet transitions of DCO$^+$ in the ν_2 state.

CHARGE DISTRIBUTION IN MOLECULAR IONS AS COMPARED WITH THOSE IN NEUTRALS

The characteristics of ionic species may be clearly manifested if one compares the charge distribution in an ion with that of isoelectronic neutrals. In the present study, a simple procedure was developed that allowed us to estimate the charge distribution in a molecule from precise structure parameters provided by high-resolution spectroscopy.

The present method is based upon Born–Oppenheimer approximation and utilizes the Hellmann–Feynman theorem, allowing us to evaluate the force acting on a nucleus in molecule. If one designates the eigenvalue of the electronic ground state as U_0, the x component of the force acting on the αth nucleus is given by (Feynman 1939)

$$-(\partial U_0/\partial X_\alpha)_e = Z_\alpha e^2 \left\{ \langle 0 \left| \sum_j (x_j - X_\alpha)/r_{j\alpha}^3 \right| 0 \rangle - \sum_{\beta \neq \alpha} Z_\beta(X_\beta - X_\alpha)/R_{\alpha\beta}^3 \right\}, \tag{7}$$

where Z denotes the nuclear charge, x_j and X_α stand for the x coordinates of the jth electron and the αth nucleus, respectively, and $r_{j\alpha}$ and $R_{\alpha\beta}$ for the distances between j and α and α and β, respectively. The first term in the right-hand side of (7) may be simplified to

$$\bar{\rho}_{x\alpha} = \int \rho(x - X_\alpha)/r_\alpha^3 \, d\tau, \tag{8}$$

where the electronic charge in a small volume $d\tau$ at r_α from the αth nucleus is expressed as $\rho \, d\tau$. We may refer to $\bar{\rho}_{x\alpha}$ as the electronic force. At the equilibrium, the electronic force is balanced by nuclear repulsion.

Application to diatomic species

Suppose the first nucleus with the charge Z_1 to be located at X_1 and the second nucleus of the charge Z_2 at X_2, with $R = X_2 - X_1$ denoting the equilibrium internuclear distance. The electronic force is given by $\bar{\rho}_{x1} = Z_2/R^2$ and $\bar{\rho}_{x2} = -Z_1/R^2$. Table 1 gives the numerical values of $\bar{\rho}$ for three isoelectronic molecules NH$^-$, OH and FH$^+$.

The NH$^-$ ion has much smaller $|\bar{\rho}|$ than the other two molecules. This may be explained by the charge distribution in NH$^-$ being more spread away from the centre of the molecule than in the other two. The last two molecules seem to have similar charge distributions, except that the charge is shifted toward the F atom in FH$^+$ a little more than toward the O atom in OH.

TABLE 1. ELECTRONIC FORCE IN DIATOMICS

	NH⁻	OH	FH⁺
$R/\text{Å}^a$	1.039^b	$0.969\,66^c$	1.0011^c
$\bar{\rho}_{x1}/\text{Å}^{-2}$	6.48	8.51	8.98
$\bar{\rho}_{x2}/\text{Å}^{-2}$	-0.93	-1.06	-1.00

^{a} $1\text{ Å} = 10^{-10}\text{ m} = 10^{-1}\text{ nm}$.
^{b} Neumark *et al.* (1985).
^{c} Huber & Herzberg (1979).

Application to linear HXY triatomics

The three atoms H, X, and Y are numbered as 1, 2 and 3 and are placed on the x-axis from left to right in this order. The electronic force is then given by

$$\bar{\rho}_{x1} = Z_2/R_{12}^2 + Z_3/R_{13}^2,$$
$$\bar{\rho}_{x2} = -Z_1/R_{12}^2 + Z_3/R_{23}^2,$$
$$\bar{\rho}_{x3} = -Z_1/R_{13}^2 - Z_2/R_{23}^2.$$

As shown in table 2, there are seven isoelectronic molecules of HXY type, of which four are positive ions and three neutral molecules. Unfortunately no negative ions of this series have been reported (examples are HBN⁻, HCC⁻, and HNB⁻).

TABLE 2. LINEAR ISOELECTRONIC HXY TRIATOMICS

(Species in parentheses have not been reported.)

X =	Be	B	C	N	O	F
Y = Be	.	.	.	.	.	(HFBe)
B	.	.	.	.	(HOB)	(HFB⁺)
C	.	.	.	HNC	HOC⁺	
N	.	.	HCN	HNN⁺		
O	.	HBO	HCO⁺			
F	(HBeF)	HBF⁺				

The electronic forces calculated for the seven molecules are plotted in figure 1, where the abscissa may be identified as $Z_X - Z_Y$. Structural data were taken from the following references: HBF⁺ (Cazzoli *et al.* 1987); HBO (Kawashima *et al.* 1987); HCO⁺ (Bogey *et al.* 1981); HCN (Winnewisser *et al.* 1971); HNN⁺ (Owrutsky *et al.* 1986); HNC (Creswell & Robiette 1978); HOC⁺ (Bogey *et al.* 1986).

It is easily seen from figure 1 that $\bar{\rho}_{x1}$ increases and $\bar{\rho}_{x3}$ decreases on going from left to right, whereas $\bar{\rho}_{x2}$ does not show any definite trend. This behaviour of $\bar{\rho}_{x1}$ and $\bar{\rho}_{x3}$ may be understood if one takes into account the electronegativities of the X and Y atoms. The second point is more important for the present problem. Both $\bar{\rho}_{1x}$ and $\bar{\rho}_{3x}$ show zigzags, namely, if one draws two curves passing through points for ions and neutrals, respectively, the one for ions is always larger in magnitude than the curve for neutrals. This means that more charge is located around the X atom in ions than in neutrals, when neighbouring species are compared. One might wonder if such zigzag behaviour is entirely ascribed to such behaviour in bond length, because the electronic force is derived from bond length. However, as shown in figure 2, this is not the case; although R_{X-Y} shows some zigzag behaviour in the central part of the figure, R_{HX} does not show any zigzags.

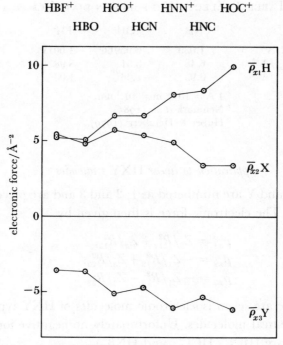

FIGURE 1. Electronic forces in linear HXY-type ions and neutrals; $\bar{\rho}_{x1}$, $\bar{\rho}_{x2}$ and $\bar{\rho}_{x3}$ represent the forces acting on H, X, and Y, respectively.

Application to bent C_{2v} XH_2-type species

Let us take the coordinate system as follows: the X atom is placed on the y-axis, whereas two hydrogens are located symmetrically about the y-axis. There are then three non-vanishing electronic forces:

$$\bar{\rho}_{x1} = -\bar{\rho}_{x2} = [Z_1/4 \sin^2\theta + Z_3 \sin\theta]/R^2,$$
$$\bar{\rho}_{y1} = \bar{\rho}_{y2} = Z_3 \cos\theta/R^2,$$
$$\bar{\rho}_{y3} = -2Z_1 \cos\theta/R^2,$$

where the atom numbers are 1 and 2 for hydrogens and 3 for X, and R and 2θ denote the X–H distance and the HXH angle, respectively. The electronic force was calculated for the third isoelectronic series NH_2^-, OH_2 and FH_2^+. Structural data were taken from the following references: NH_2^- (Tack *et al.* 1986); H_2O (Callomon *et al.* 1976); H_2F^+ (Schäfer & Saykally 1984). It is interesting to note that the angle $2 \arctan(\bar{\rho}_{y1}/\bar{\rho}_{x1})$ is larger than 2θ by 4.1°, 3.8°, and 3.0°, respectively, for the three molecules. The electronic force $[\bar{\rho}_{x1}^2 + \bar{\rho}_{y1}^2]^{\frac{1}{2}}$ acting on the hydrogen is compared with $\bar{\rho}_{x1}$ of the diatomics above mentioned in figure 3, where a similar comparison is also made for $\bar{\rho}_{y3}$ of XH_2 and $\bar{\rho}_{x2}$ of the diatomics. The similarities are very striking.

LCAO MO interpretation of the electronic force

It may be instructive to calculate the electronic force by using an LCAO MO (linear combination of atomic orbitals molecular orbital) for the wavefunction. Two atoms or nuclei called A and B are placed on the x-axis with the separation $R = X_B - X_A$. The LCAO MO is expressed

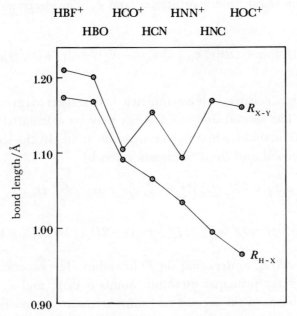

FIGURE 2. Bond lengths in linear HXY-type ions and neutrals.

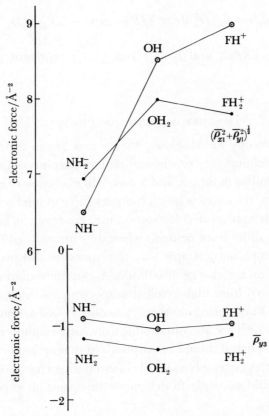

FIGURE 3. Electronic forces in XH-type diatomics and C_{2v} H_2X-type triatomics. The upper traces apply to hydrogens and the lower traces to X atoms.

as $\psi = c_A u_A + c_B u_B$ in terms of the atomic orbitals u_A and u_B. The electronic force is then given by

$$\bar{\rho}_{xA} = |c_A|^2 \int |u_A|^2 (x - X_A)/r_A^3 \, d\tau + 2Re[c_A^* c_B \int u_A^* u_B (x - X_A)/r_A^3 \, d\tau] + |c_B|^2 \int |u_B|^2 (x - X_A)/r_A^3 \, d\tau.$$
$$(9)$$

The first term of (9) is zero, because $|u_A|^2$ is symmetric with respect to inversion at X_A, whereas $x - X_A$ is antisymmetric. The second and third terms may be calculated analytically, if the atomic orbitals are approximated by hydrogenic wavefunctions. In the simplest case of both u_A and u_B being 1s, the second and third terms are given by

$$\int u_A u_B (x - X_A)/r_A^3 \, d\tau = (Z/a_0)^2 \exp(-D_1) D_1$$

and
$$\int u_B^2 (x - X_A)/r_A^3 \, d\tau = (Z/a_0)^2 [1/D_1^2 - \exp(-2D_1)(2 + 2/D_1 + 1/D_1^2)],$$

respectively, where $D_1 = ZR/a_0$, a_0 denoting the Bohr radius. The expressions are quite complicated if $Z_A \neq Z_B$ and if the principal quantum numbers of u_A and u_B are larger than 1. However, the functional forms of the second and third terms are proportional to

$$\int u_A u_B (x - X_A)/r_A^3 \, d\tau \propto \exp(-D_n) f_2(D_n)$$

and
$$\int u_B^2 (x - X_A)/r_A^3 \, d\tau \propto 1/R^2 - \exp(-2D_n) f_3(D_n),$$

respectively, where $D_n = ZR/na_0$ and $f_2(D_n)$ and $f_3(D_n)$ represent polynomial functions of D_n.

DISCUSSION AND CONCLUSION

High-resolution spectroscopy has disclosed that a new group of molecules, i.e. molecular ions, exist and has been yielding very precise molecular constants of these species. Many more interesting examples will follow in future, and a new field of chemistry is expected to be opened. Because of their net charge, the ions are much more sensitive to interactions with environments than neutral molecules, as appreciated for some time. However, it has not been explored in detail how molecular ions differ from neutrals when they are isolated from environments. The present study aimed at proposing a simple way to respond to this problem. The information derived from this method on the charge distribution is not more than qualitative, but is based upon structural data derived from high-resolution spectroscopic experiments, which are very reliable. The method may be extended to utilize harmonic as well as anharmonic force constants and hyperfine constants, which are also obtainable from high-resolution data (see, for example, Salem 1963). In this connection it should be emphasized that an important molecular constant of ions, the dipole moment, is extremely valuable in estimating charge distribution in molecular ions, and much effort should be made to determine this constant experimentally.

REFERENCES

Amano, T., Kawaguchi, K. & Hirota, E. 1987 *J. molec. Spectrosc.* (In the press.)

Bogey, M., Demuynck, C. & Destombes, J. L. 1981 *Molec. Phys.* **43**, 1043–1050.

Bogey, M., Demuynck, C. & Destombes, J. L. 1986 *J. molec. Spectrosc.* **115**, 229–231.

Callomon, J. H., Hirota, E., Kuchitsu, K., Lafferty, W. J., Maki, A. G. & Pote, C. S. 1976 *Landolt–Börnstein tables, new series, group II*, vol. 7. Springer: Heidelberg.

Cazzoli, G., Degli Esposti, C., Dore, L. & Favero, P. G. 1986 *J. molec. Spectrosc.* **119**, 467.

Cazzoli, G., Degli Esposti, C., Dore, L. & Favero, P. G. 1987 *J. molec. Spectrosc.* **121**, 278–282.

Creswell, R. A. & Robiette, A. G. 1978 *Molec. Phys.* **36**, 869–876.

De Lucia, F. C. & Helminger, P. A. 1977 *J. chem. Phys.* **67**, 4262–4267.

Feynman, R. P. 1939 *Phys. Rev.* **56**, 340–343.

Hirota, E. & Endo, Y. 1987 *J. molec. Spectrosc.* (In the press.)

Huber, K. P. & Herzberg, G. 1979 *Molecular spectra and molecular structure*, vol. 4. *Constants of diatomic molecules.* New York: Van Nostrand.

Kawaguchi, K. & Hirota, E. 1986*a* *Chem. Phys. Lett.* **123**, 1–3.

Kawaguchi, K. & Hirota, E. 1986*b* *J. chem. Phys.* **84**, 2953–2960.

Kawaguchi, K. & Hirota, E. 1986*c* *J. chem. Phys.* **85**, 6910–6913.

Kawaguchi, K., McKellar, A. R. W. & Hirota, E. 1986 *J. chem. Phys.* **84**, 1146–1148.

Kawashima, Y., Endo, Y., Kawaguchi, K. & Hirota, E. 1987 *Chem. Phys. Lett.* **135**, 441–445.

Owrutsky, J. C., Gudeman, C. S., Martner, C. C., Tack, L. M., Rosenbaum, N. H. & Saykally, R. J. 1986 *J. chem. Phys.* **84**, 605–617.

Neumark, D. M., Lykke, K. R., Andersen, T. & Lineberger, W. C. 1985 *J. chem. Phys.* **83**, 4364–4373.

Saito, S., Yamamoto, S. & Kawaguchi, K. 1987 *J. chem. Phys.* **86**, 2597–2599.

Salem, L. 1963 *J. chem. Phys.* **38**, 1227–1236.

Sastry, K. V. L. N., Herbst, E. & De Lucia, F. C. 1981 *J. chem. Phys.* **75**, 4169–4170.

Schäfer, E. & Saykally, R. J. 1984 *J. chem. Phys.* **81**, 4189–4199.

Tack, L. M., Rosenbaum, N. H., Owrutsky, J. C. & Saykally, R. J. 1986 *J. chem. Phys.* **85**, 4222–4227.

Winnewisser, G., Maki, A. G. & Johnson, D. R. 1971 *J. molec. Spectrosc.* **39**, 149–158.

Note added in proof (8 *September* 1987). Kawaguchi & Hirota (submitted to *J. chem. Phys.*) have recently detected the ν_3 band of the FHF$^-$ ion at 1331.1507 (7) cm^{-1}, together with the ν_2 (1286.0284 (22) cm^{-1}) and $\nu_1 = \nu_3 - \nu_1$ (1265.6450 (19) cm^{-1}) bands. The former ν_3 band was reassigned to $\nu_1 + \nu_3$.

Phil. Trans. R. Soc. Lond. A **324**, 141–146 (1988)

Printed in Great Britain

Microwave spectroscopy of molecular ions in the laboratory and in space

By R. C. Woods

Department of Chemistry, University of Wisconsin, Madison, Wisconsin 53706, U.S.A.

The SO^+ molecular ion has been detected radioastronomically via the $J = \frac{5}{2} \to \frac{3}{2}$ rotational transitions, whose rest frequencies were determined by earlier laboratory spectroscopic studies. This ion was detected in seven interstellar sources, including both giant molecular clouds and a cold dark cloud, and thus appears to be very widely distributed in the Galaxy and to play an important role in interstellar chemistry. We have obtained rotational spectra of HCO^+ in a wide variety of vibrational states in three isotopic forms, leading to the equilibrium structural parameters that are consistent to a high degree in redundant determinations. These results will be compared to similar structures we have obtained for HCN and HNC. The high bending vibrational states observed for HCO^+ and HCN exhibit the effects of Stark broadening due to the electric fields present in the discharge plasmas. The effects observed for ions against neutrals will be compared, as will those for normal glow discharges against those for magnetically confined abnormal discharges.

INTRODUCTION

The applications of microwave spectroscopy to the study of molecular ions fall into three fairly distinct categories: precise determinations of geometrical and electronic structures of the ionic species, detection of ions in interstellar space by radioastronomy, and plasma diagnostics, i.e. probes of the ion's dynamical interaction with its environment. Because microwave transition frequencies can be measured with great accuracy and because they depend very directly on the moments of inertia of the species and thus its bond distances and angles, very accurate molecular structural information is potentially available from the technique. To fully exploit this potential, however, one must be able to obtain data in a sufficient number of excited vibrational states so that the effects of vibration–rotation interaction may be fully accounted for. Of all the bands of the electromagnetic spectrum the microwave region has proven especially suitable for detection of ions and other molecules in interstellar space, for two reasons. The regions where molecular species exist are very cold (so that the only upper states sufficiently excited to emit many photons are those of rotational transitions) and always associated with large amounts of dust (so that they are highly obscured or totally opaque to shorter-wavelength radiation but still transparent to microwave photons). In the laboratory plasma environment the shapes, widths, relative intensities and frequency displacements of microwave spectral transitions (as well as of those in other forms of high-resolution spectroscopy) can provide a great deal of information about the distribution of energy between vibration, rotation and translation in various species, and in the case of ions about the ion velocities, electric-field strengths, and plasma densities. The present discussion is in no way intended to be a comprehensive review of these areas, but rather will describe one topic in my recent research experience illustrating current efforts in each of the three categories.

Radioastronomy of SO$^+$

In early 1985, our group was able to obtain a laboratory microwave spectrum of the SO$^+$ molecular ion (Warner *et al.* 1987) in its ground $^2\Pi_{\frac{1}{2}}$ state. We were able to measure five transitions that were sufficient to determine the effective rotational, centrifugal distortion, and lambda doubling constants and thus to predict any other rotational transitions in this $\Omega = \frac{1}{2}$ substate. Although the transitions were weak and there were numerous stronger nearby lines in any of the gas mixtures we used, the SO$^+$ transitions could be unambiguously identified by the very characteristic slow but observable Zeeman effect of a $^2\Pi_{\frac{1}{2}}$ state. The availability of this new laboratory data then introduced the possibility of a serious radioastronomical search for SO$^+$; this seemed particularly worthwhile because the ion–molecule models, e.g. that of Herbst & Leung (1986), predicted that SO$^+$ was one of the more abundant molecular ions in the interstellar medium, among those that had not been previously observed.

Thus the author entered into a collaboration with radioastronomers E. B. Churchwell, W. M. Irvine, and R. L. Dickman with the goal of detecting interstellar SO$^+$. This has been accomplished (Woods *et al.* 1987) in a series of four observing runs: three at the National Radio Astronomy Observatory, Kitt Peak, Arizona, and one at the Five College Radio Astronomy Observatory near Amherst, Massachusetts. The first three runs involved exclusively efforts on the $J = \frac{5}{2} \rightarrow \frac{3}{2}$ lambda doublet at 115.8 GHz and 116.2 GHz. Both of these doublet components were detected in six giant molecular clouds: SGRB2, ORION-A, NGC-7538, NGC-2264, DR21-OH, and W51. In figure 1 the resulting spectra obtained in the source DR21-OH are

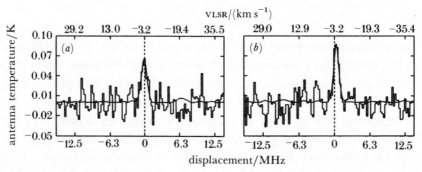

FIGURE 1. The $J = \frac{5}{2} \rightarrow \frac{3}{2}$ transitions of SO$^+$ in the giant molecular cloud DR21-OH. The smooth curve overlaying the spectrum of raw data is from the statistical analysis of Brown *et al.* (1985). (*a*) 115.8 GHz line; (*b*) 116.2 GHz line.

shown. Each spectrum involves an average of data from two different runs at NRAO (September 1985 and June 1986) with two different receivers. The figure also shows statistically smoothed curves based on a method of Brown *et al.* (1985) superimposed on the histogram plot of the raw data. This smoothing procedure has proved to be very useful in the SO$^+$ studies. In each of the six sources the intensities, Doppler shifts, and linewidths of the two lines were as similar as could be expected with the modest available signal:noise ratio. The actual values of the velocity shifts were also as expected for the individual sources. In addition we have barely detected one of the doublet components in the cold dark cloud L183.

All this observational information taken together, we believe, provides a sufficient basis to conclude that the observed features are indeed caused by SO$^+$. Somewhat surprisingly, the intensities of the lines are comparable (0.05–0.10 K) in all of the seven sources. Detecting lines of this strength with currently available receivers requires several hours of signal averaging for each spectrum. Because we have detected signals for SO^+ in all seven of the sources in which we have looked, it appears that SO^+ is widely distributed in molecular clouds in the galaxy. To determine the overall abundance of SO^+ in the various sources reliably one must measure the degree of excitation, i.e. the rotational temperature, by measuring transitions with at least one more J value. The observed $J = \frac{5}{2} \rightarrow \frac{3}{2}$ emission intensities are sensitive only to the population in the $J = \frac{5}{2}$ rotational level. With this in mind our most recent observing run at NRAO (March 1987) involved efforts to observe the $J = \frac{11}{2} \rightarrow \frac{9}{2}$ transitions of SO^+ near 255 GHz. It turns out that these frequencies are more plagued with interfering lines than the earlier ones, and the results must certainly be considered preliminary at this point. We do appear to have detected SO^+ in NGC-7538, although very weakly. It appears that the $J = \frac{11}{2} \rightarrow \frac{9}{2}$ lines are quite weak and thus that the excitation temperatures are fairly low (less than or equal to about 20 K). Yet another series of observations is planned in the near future with the goal of further refining our estimate of the excitation. When overall SO^+ abundances are observationally available they can be compared to available theoretical models, and this will hopefully lead to further refinement of the latter and to a better understanding of interstellar sulphur chemistry. The SO^+ ion is only the second sulphur-containing molecular ion (after HCS^+ (Thaddeus *et al.* 1981)) to be observed in the interstellar medium. (It is also the only radical ion so far observed.)

EQUILIBRIUM STRUCTURE OF HCO^+

Substitution type (r_s) structures have been available for several years for HCO^+ (Woods *et al.* 1981), HNN^+ (Szanto *et al.* 1981), HOC^+ (Gudeman & Woods 1982) and the related neutral species HNC and HCN (Pearson *et al.* 1976). These structures are obtained from ground-vibrational-state data for the rotational constants or moments of inertia in several isotopic forms by using a simple well-known procedure (Kraitchman 1953). A more fundamental type of structure is the equilibrium (r_e) type, which gives the location of the actual minimum in the Born–Oppenheimer potential. It is the r_es that are predicted by *ab initio* quantum chemical calculations, and because the discrepancies of a few thousandths of an ångström (Å)† between r_e and r_s structures are comparable in magnitude to the uncertainties in the best *ab initio* structure predictions, an experimental r_e structure is essential for a full calibration of the accuracy of such a calculation. Experimentally r_es are gleaned from B_es that are determined from the effective rotational constants for each vibrational state via the equation

$$B_{v_1 v_2{}^l v_3} = B_e - \alpha_1(v_1 + \tfrac{1}{2}) - \alpha_2(v_2 + 1) - \alpha_3(v_3 + \tfrac{1}{2}) + \gamma_{11}(v_1 + \tfrac{1}{2})^2 + \gamma_{22}(v_2 + 1)^2 + \gamma_{33}(v_3 + \tfrac{1}{2})^2$$

$$+ \gamma_{12}(v_1 + \tfrac{1}{2})(v_2 + 1) + \gamma_{13}(v_1 + \tfrac{1}{2})(v_3 + \tfrac{1}{2}) + \gamma_{23}(v_2 + 1)(v_3 + \tfrac{1}{2}) + \gamma_{ll} l^2. \quad (1)$$

For simplicity the effects of l-type doubling and l-type resonance have been omitted from this equation, but they are fully accounted for in our analysis. At a minimum spectra in the ground vibrational state and one excited state of each of the three normal modes must be assigned. Then one can solve for B_e and the three αs with the γ parameters arbitrarily set to zero. We did this for HCO^+ several years ago (Gudeman 1982) by obtaining the (100), (001), and (02^{0}0) satellite spectra for the normal isotopic species and also for $H^{13}CO^+$ and $HC^{18}O^+$.

† $1 \text{ Å} = 10^{-1} \text{ nm} = 10^{-10} \text{ m}.$

Because there are only two bond distances, they can be determined independently by using any of the three possible isotope pairs. The results were disappointing because the three different values obtained for each bond length were not at all consistent, varying by as much as 0.01 Å. Considerable speculation on the origin of this inconsistency ensued. Exactly equivalent data were obtained for the isoelectronic molecule HCN, but in that case the three structure determinations appeared beautifully consistent. Was an ion in some critical way different from a neutral species? Was there an error in the HCO^+ data? Was there a perturbation between vibrational states in one of the isotopes? The last possibility eventually seemed most likely, particularly in the form of a possible third-order anharmonic resonance interaction between the observed (100) state and the (04^00) state.

We have now satisfactorily resolved these questions by obtaining a much more extensive set of vibrational satellite data for the three isotopic forms of HCO^+ (Gudeman et al. 1987). We have observed all three in the (100), (001), (01^10), (02^00), (02^20), (03^10), (03^30), (04^00), (002) and (01^11) states, so that γ_{22}, γ_{33}, γ_{23} and γ_{ll}, and an even higher-order parameter ϵ_{222}, can be determined in addition to B_e and the αs. In each state, centrifugal distortion parameters were measured. These are required to carry out the analysis of l-type resonance (DeLucia & Helminger 1977) mentioned earlier. Because the (04^00) satellites are found exactly where they are predicted from extrapolating the pattern of the lower bending states, we have eliminated the possibility of the suspected $(100)-(04^00)$ interaction in all three isotopic forms. In fact the earlier problems arose from the neglect of all the γ parameters in the previous treatment. Some of the γs are more isotope dependent in HCO^+ than they are in HCN. We now have B_es (taking into account αs but not γs) for HNC and the r_e structure in this case shows some of the same inconsistency problems that HCO^+ did. It seems now that HCN is somewhat fortuitous in working so well when only the α terms are used. The CO distance for HCO^+ is now consistent to 0.00002 Å and the CH distance to 0.0001 Å. A summary of r_e and r_s structures for HCO^+, HCN, HNC and HNN^+ (with the r_e structure of HNN^+ taken from the infrared work of Owrutsky et al. 1986) is given in table 1, where it can be seen that the $r_e - r_s$ values are fairly similar across this set of molecules.

TABLE 1. COMPARISON OF r_e AND r_s FOR ISOELECTRONIC SERIES

(Work at University of Wisconsin, except r_e for HNN^+. Numbers in parentheses indicate only the scatter (1σ) between determinations from various isotopic combinations. Realistic errors including systematic effects are larger.)

		$r(XY)/\text{Å}$	$r(XH)/\text{Å}$
HCO^+	r_e	1.104738 (23)	1.097247 (38)
	r_s	1.107211 (15)	1.092881 (35)
	$r_e - r_s$	0.0025	0.0044
HCN	r_e	1.153193 (16)	1.065825 (103)
	r_s	1.155461 (25)	1.063091 (33)
	$r_e - r_s$	0.0023	0.0027
HNC	r_e	1.168363 (226)	0.996959 (1450)
	r_s	1.172055 (7)	0.985884 (120)
	$r_e - r_s$	0.0037	0.0111
HNN^+	r_e[a]	1.097266 (92)	1.03359 (43)
	r_s	1.095415 (6)	1.031426 (56)
	$r_e - r_s$	0.0026	0.0022

[a] Data from Owrutsky et al. (1986). Errors in parentheses are quoted from that source.

STARK EFFECTS IN PLASMAS

The same vibrational satellite spectra of HCO^+ that have served to resolve the questions about the equilibrium structure have also provided some important new information on a dynamical property of the plasma, namely the prevailing electric-field strengths. Most microwave lines of molecular ions have rather slow Stark effects, so that the presence of the electric fields that are typically present in plasmas has a negligible impact on their appearance. Transitions involving degenerate or near-degenerate energy levels, e.g. vibrational satellites of linear molecules with high values of the bending quantum number v_2 and the vibrational angular-momentum quantum number l, however, have much faster Stark effects. Thus we have observed (in collaboration with G. Cazzoli) clear effects of the Stark perturbation in the (01^10), (02^20) and (03^30) states of HCO^+ (Conner et al. 1987). These include shifts of centre frequency, dramatic changes in peak amplitude, and increases in linewidth. We have carried out measurements in both normal DC glow discharges and in magnetically confined, abnormal glows, with magnetic fields of approximately 300 G† Furthermore, we have looked at the corresponding transitions in HCN in both types of discharge. (It is of course of interest to know if an ion and a neutral species experience the same electric field or not.)

Some preliminary results may be stated. The electric fields involved are not the macroscopic ones, but the microscopic fields from the local distribution of charged particles in the plasma. The macroscopic fields are entirely too small to explain the observed effects, especially in the abnormal discharge case, where the macroscopic fields have been shown to be very small. We have used the simple Holtsmark theory (Holtsmark 1919; Griem 1974) to estimate the distribution of electric fields for different plasma densities, and then used this along with a quasistatic approximation to simulate the observed spectral behaviour of the (02^20) or (01^10) transition. The observed effects can be at least qualitatively explained in this way, except in the case of an ion in the magnetically confined plasma. In that case, the effect is distinctly bigger than for the corresponding neutral, for reasons that are not yet understood. The (03^30) state presents a somewhat different case. Here the degeneracy is very much closer, so that the Stark effect is first order, rather than second. We find this satellite to be perturbed, but not as strongly as would be expected from a simulation of the above type (that would predict total obliteration for a line with such a fast Stark effect). This behaviour is probably caused by a breakdown in the validity of the quasistatic approximation because the frequencies of the fluctuations in the field are greater than the frequency separation betwen the interacting levels.

This work was supported by the Physical Chemistry Program of the National Science Foundation, the donors of the Petroleum Research Fund, administered by the American Chemical Society, and by the Wisconsin Alumni Research Foundation.

REFERENCES

Brown, R. D., Godfrey, P. D. & Rice, E. H. N. 1985 Observatory 105, 12–15.
Conner, W. T., Warner, H. E., Woods, R. C. & Cazzoli, G. 1987. (In preparation.)
DeLucia, F. C. & Helminger, P. A. 1977 J. chem. Phys. 67, 4262–4267.
Griem, H. R. 1974 Spectral line broadening by plasmas, Chs 1 & 2. New York: Academic Press Inc.
Gudeman, C. S., Warner, H. E., Woods, R. C. & Cazzoli, G. 1987 (In preparation.)

† $1 G = 10^{-4} T$.

Gudeman, C. S. 1982 Doctorial dissertation, University of Wisconsin.

Gudeman, C. S. & Woods, R. C. 1982 *Phys. Rev. Lett.* **48**, 1344–1348.

Herbst, E & Leung, C. M. 1986 *Mon. Not. R. astr. Soc.* **222**, 689–710.

Holtsmark, J. 1919 *Annln Phys.* **58**, 577–587.

Kraitchman, J. 1953 *Am. J. Phys.* **21**, 17–24.

Owrutsky, J. C., Gudeman, C. S., Martner, C. C., Tack, L. M., Rosenbaum, N. H. & Saykally, R. J. 1986 *J. chem. Phys.* **84**, 605–617.

Pearson, E. F., Creswell, R. A., Winnewisser, M. & Winnewisser, G. 1976 *Z. Naturf.* A **31**, 1394–1397.

Szanto, P. G., Anderson, T. G., Saykally, R. J., Piltch, N. D., Dizon, T. A. & Woods, R. C. 1981 *J. chem. Phys.* **75**, 4261–4263.

Thaddeus, P., Guélin, M. & Linke, R. A. 1981 *Astrophys. J.* **246**, L41–46.

Warner, H. E., Carballo, N. & Woods, R. C. 1987 (In preparation.)

Woods, R. C., Churchwell, E. B., Irvine, W. M. & Dickman, R. L. 1987 (In preparation.)

Woods, R. C., Saykally, R. J., Anderson, T. G., Dixon, T. A. & Szanto, P. G. 1981 *J. chem. Phys.* **75**, 4256–4260.

Discussion

R. J. SAYKALLY (*University of California, Berkeley, U.S.A.*). Professor Woods has reported the equilibrium structure of HCO^+ determined with incredibly high precision. It becomes questionable what the physical significance of such high precision structures really is, because subtle effects of electronic contributions to the moments of inertia become important at the 0.001 Å level. These require knowledge of both the rotational g-factor and the molecular potential surface to correctly account for them. At present this knowledge is not available for HCO^+, although we have recently published results for ArH^+ in which these corrections were made to obtain the adiabatic internuclear separation.

Reference

Laughlin, K. B., Blake, G. A., Cohen, R. C., Hovde, D. C. & Saykally, R. C. 1987 *Phys. Rev. Lett.* **58**, 996–999.

R. C. WOOD. There are certainly some interesting theoretical problems involved in the determination of highly accurate molecular structures for polyatomic molecules, including some that are specific to ionized species. We cannot claim to have solved all these problems, but at least there are now some data of sufficient precision to make their consideration worthwhile. We are continuing to work on these details of the interpretation of the spectral data in terms of structure.

Phil. Trans. R. Soc. Lond. A **324**, 147–162 (1988)

Printed in Great Britain

Millimetre-wave and submillimetre-wave spectroscopy of molecular ions

By J. L. Destombes, C. Demuynck and M. Bogey

*Laboratoire de Spectroscopie Hertzienne, Université des Sciences et Techniques de Lille,
59655 Villeneuve D'Ascq Cedex, France*

In the last few years, the specificities of millimetre-wave spectroscopy coupled with active developments in other fields (*ab initio* calculation, radioastronomy and infrared spectroscopy) have led to an impressive increase in our knowledge of the spectral characteristics of molecular ions. The advances to be reviewed in this paper are a result of the development of highly sensitive spectrometers and of efficient methods of production. Some specific examples, including ions with large-amplitude motion (H_3O^+), weakly polar ions ($HCNH^+$), asymmetric-top ions (HCO_2^+, HON_2^+) and ionic clusters ($Ar \cdot H_3^+$), will be used to illustrate the potentialities of millimetre- and submillimetre-wave spectroscopy.

1. Introduction

The exciting story of high-resolution rotational spectroscopy of molecular ions began in 1970 with the discovery of the so-called X-ogen interstellar line by Buhl & Snyder (1970) and the subsequent proposal by Klemperer (1970) that this line was caused by the HCO^+ ion. This identification, strongly supported by *ab initio* calculations (Wahlgreen *et al.* 1973; Kramers & Diercksen 1976) and by the observation of an interstellar line attributed to the ^{13}C substitution (Snyder *et al.* 1976), became definitive with the laboratory detection of HCO^+ in a glow discharge by Woods & co-workers (Woods *et al.* 1975). HCO^+ has been the first example of a remarkably fruitful approach with radioastronomical results, *ab initio* calculations and laboratory measurements. HN_2^+, HCS^+, HCO_2^+ have been also first observed in the interstellar medium (Turner 1974; Green *et al.* 1974; Thaddeus *et al.* 1981) before being detected in the laboratory (Saykally *et al.* 1976a; Gudeman *et al.* 1981; Bogey *et al.* 1984b). The discovery of the key role played by ions in interstellar chemistry has been indeed very stimulating for radioastronomers, theoreticians, and laboratory spectroscopists: more than 40 ions are now characterized by their millimetre spectra and/or infrared (IR) spectra.

In addition to its significance for astrophysics, millimetre-wave spectroscopy also gives essential information about energy-level configurations and molecular structure. Millimetre-wave spectroscopy, because of its frequency range, is inherently a very high-resolution technique (typically of the order of 100 kHz) and makes possible a detailed analysis of the rotational, fine and hyperfine structures and leads to an accurate determination of the molecular parameters.

During the last few years, these advantages have been shared with tunable IR spectroscopy. As these aspects are described in detail in other papers at this symposium, we will restrict ourselves to the methods and contributions of millimetre- and submillimetre-wave spectroscopy with tunable sources. We will consider first the highly sensitive spectrometers needed to observe molecular ions and then the efficient methods of ion production developed in different groups. Finally, we will illustrate the power of millimetre-wave spectroscopy by describing some recent results we obtained in Lille.

2. Millimetre and submillimetre spectrometers

With the exception of a few promising experiments with molecular beams (Brown *et al.* 1981; Johnson *et al.* 1984) that will not be described here, molecular ions are generally produced and observed inside low-pressure plasmas. Even with optimized methods of production, the ion density is always very weak, and the main quality required of a spectrometer is a detection threshold as low as possible. Considering a linear molecule as an example, the lowest detectable concentration N_{min} (per cubic centimetre) is given by the following expression:

$$N_{min} \approx 6.56 \times 10^{27} \frac{T^2 \Delta \nu \gamma_{min}}{\mu^2 \nu_0^3} e^{W/0.69T},$$

where T is the temperature of the gas, assuming thermal equilibrium, in kelvins; $\Delta \nu$ is the half width at half maximum, in megahertz; γ_{min} is the minimum detectable absorption coefficient, in reciprocal centimetres; μ is the dipole moment, in debyes†; ν_0 is the frequency of the line, in megahertz; and W is the energy of the lower level of the transition, in reciprocal centimetres.

As illustrated in table 1, it is clear that two parameters can play a favourable role.

TABLE 1. LOWEST DETECTABLE CONCENTRATION, N_{min}, CALCULATED FOR SOME TYPICAL CONDITIONS ASSUMING A SPECTROMETER SENSITIVITY OF 10^{-7} CM^{-1}

μ/D	ν/GHz	T/K	N_{min}/cm^{-3}	example
1.8	128	300	3×10^9	HCS$^+$
3.8	90	100	4×10^8	HCO$^+$
3.8	360	100	8×10^6	
0.5	370	200	7×10^8	HCNH$^+$

1. Cooling the cell increases the absorption coefficient and has often proved to be very effective (sometimes necessary) (Dixons & Woods 1975). Nevertheless, when condensable chemicals are involved in the gas mixture, only moderate cooling can be used, and no gain is to be expected from this parameter (Gudeman *et al.* 1981; Bogey *et al.* 1985a).

2. Working in a higher-frequency range permits the detection of weaker ionic concentrations, provided that the sensitivity of the spectrometer is the same as at lower frequencies. In the millimetre range (*ca.* 100–150 GHz) a typical value $\gamma_{min} \approx 10^{-7}$ cm^{-1} can be obtained for the lowest detectable absorption coefficient in a routine experiment. This moderate sensitivity is mainly because of specific constraints related to discharges such as relatively short cells and discharge instabilities. Such a sensitivity is nevertheless difficult to reach in the submillimetre-wave range if harmonic generation from low-frequency klystrons is used, because of the lower power available. But with a relatively powerful radiation source, detection thresholds as low as those shown in table 1 can be achieved and the detection of minor ionic species in the discharge is possible.

In the millimetre-wave range, fundamental klystrons are currently available up to about 120 GHz. Although klystrons are also used at higher frequencies (Saito *et al.* 1985), generally, radiation in the 120–300 GHz range is obtained by harmonic generation by using point contact or Schottky barrier diodes. Depending on the frequency range, point contact, Schottky barrier diodes or liquid-helium-colled InSb bolometers have been used as detectors.

† $1 D \approx 3.33 \times 10^{-30}$ C m.

The spectrometer developed in Lille is of conventional design and has been described elsewhere (Bogey *et al.* 1986*c*). Varian klystrons ($f < 80$ GHz) are phase locked to a solid-state oscillator driven by a synthesizer (ADRET 6100 B). A 12.5 kHz frequency modulation is applied to the klystron by modulating the reference frequency of the phase-lock synchronizer. Above 120 GHz, the millimetre-wave power is obtained by harmonic generation by using Schottky barrier diodes mounted in commercial Custom Microwave multipliers, and detection is achieved with an InSb detector (IRD4 Advanced Kinetics) operating at 4 K. After synchronous detection, the signal is processed by a microprocessor system that ensures, in addition to the control of the frequency scans, multichannel averaging, base-line suppression, signal smoothing and line-centre frequency measurement, the later being achieved by taking the mean value of increasing and decreasing frequency scans (Bogey *et al.* 1986*c*).

In the submillimetre-range, harmonic generation is often used (Sastry *et al.* 1981; Verma *et al.* 1985; Bogey *et al.* 1986*f*). For the study of unstable species, which requires high sensitivity, this technique is generally limited to about 500 GHz, and other methods have been developed for frequencies in the terahertz region or above. Tunable far-infrared radiation has been generated by mixing the radiation of submillimetre-wave lasers with that of a klystron by using Schottky barrier diodes (Van den Heuvel *et al.* 1980; Blake *et al.* 1986; Laughlin *et al.* 1986), or by mixing two CO_2 laser lines in metal–insulator–metal diodes (Evenson *et al.* 1984).

In Lille, we use a backward wave oscillator (Thomson C.S.F. TH 4218 D) that delivers at least 10 mW in the 330–406 GHz frequency range.

The short-term frequency stability is adequate to use the carcinotron in a free running mode. Figure 1*a* presents a recording of the 1_{11}–1_{10} transition of H_2D^+ (Bogey *et al.* 1984*a*) with this technique. This is a considerable bonus when searching for a new molecule with a large uncertainty on the line frequencies because, in these conditions, a very large frequency range can easily be scanned (up to 40 GHz continuously, such as in the case of $Ar \cdot D_3^+$, to be discussed

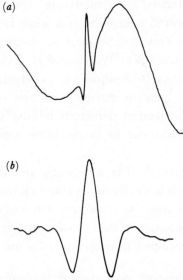

FIGURE 1. Two examples of the use of the backward wave oscillator in the submillimetre-wave range. (*a*) Free-running mode for the 1_{11}–1_{10} H_2D^+ line near 372 GHz. Lock-in time constant = 100 ms; the width of the region shown is 16 MHz and the sweeping time is about 30 s. The crooked base line is because of standing waves in the cell. (*b*) Phase-locked mode for the $J = 5 \leftarrow 4$ transition of $HCNH^+$ at about 370 GHz. Lock-in time constant = 10 ms; the signal has been averaged 120 times.

later). When signal averaging is required, the carcinotron is phase locked (Bogey *et al.* 1986*b*). A part of the submillimetre radiation is extracted with a beam splitter and mixed on a Schottky diode with the harmonics of phase-locked millimetre-wave klystron radiation. The microprocessor system that drives the millimetre-wave spectrometer described above can then be used to control the submillimetre spectrometer, leading to a significant improvement in the sensitivity. This technique allows the detection of very weak signals arising, for example, from weakly polar molecules such as $HCNH^+$, as illustrated in figure 1*b*.

As a check of the sensitivity of the spectrometer the $J = 30 \leftarrow 29$ transition of OCS in the $(10^0\ 1)$ vibrational state has been observed at $361\,572.17$ MHz with a signal:noise ratio of two in a single scan with a lock-in time constant of 30 ms, corresponding to a minimum detectable absorption coefficient of 5×10^{-8} cm^{-1} in this frequency range.

3. Methods of production

In spite of the unrecognized pioneering work of Low & Ramberg (1955), who in 1955 observed OH Λ doubling transitions inside a DC discharge, the noise generated by the plasma electrons had been thought for a long time to prevent the possibility of observing millimetre-wave spectra in an internal discharge. In 1974, however, Woods & co-workers demonstrated that this technique was very promising by observing some transient molecules inside a DC glow discharge (Woods & Dixon 1974). They succeeded in detecting the first laboratory rotational spectrum of a molecular ion, CO^+ (Dixon & Woods 1975).

The characteristics of a glow discharge depend on various parameters such as the nature and pressure of the gas, and the shape of the electrodes (Badareu & Popescu 1968; Von Engel 1965). In fact, the discharge is not spatially homogeneous and two parts of a glow discharge have characteristics compatible with the requirements of millimetre-wave spectroscopy, i.e. mainly a low and homogeneous electric field.

1. The *positive column* is characterized by a moderate (less than 10 V cm^{-1}) and uniform electric field. This macroscopic electric field induces a weak Doppler shift on the absorption frequencies of the molecular ions and can be used to discriminate these lines from those originating from neutrals (Gudeman 1982; Woods 1983). This phenomenon is the basis of the very efficient velocity-modulation technique extensively used in the IR spectral range (Gudeman & Saykally 1984), in which the direct interaction between the radiation and the plasma is negligible. Although the electron density is relatively low (*ca.* 10^9–10^{10} cm^{-3}), the column density in the positive column can be rather large because it fills the largest part of the discharge tube.

2. In the *negative glow*, the electric field is very weak and the Doppler shift is negligible (Bowman *et al.* 1982). Moreover, the ionic density is about one order of magnitude higher than in the positive column, making the negative glow very attractive for absorption spectroscopy. However, in normal conditions, the length of the negative glow is very small (a few centimetres) and special devices have to be developed to lengthen it for use in absorption spectroscopy.

3.1. *Millimetre-wave spectroscopy in the positive column*

The earliest microwave observations inside a low-pressure plasma took place in the positive column of a DC discharge (Low & Ramberg 1955; Woods & Dixon 1974; Dixon & Woods 1975) and a number of molecular ions have now been detected by this technique (Sastry *et al.*

1981; Woods 1983). This technique is also very efficient for the production of free radicals (Dixon & Woods 1977; Hirota 1985), metastable species (Saykally *et al.* 1976*b*; Saykally 1977), unstable molecules (Thaddeus *et al.* 1986; Tang *et al.* 1985) as well as vibrationally excited species (Saykally 1977; Skatrud *et al.* 1983; Gudeman 1982). Although the positive column has been used very successfully in the study of elusive molecules, some problems occasionally arise from the presence of metal electrodes inside a very reactive plasma (Woods & Dixon 1974; Saykally *et al.* 1976*a*). In Lille we have developed an electrodeless radio-frequency discharge that avoids these problems (Bustreel *et al.* 1979; Bogey *et al.* 1981) and has plasma characteristics very similar to those of a positive column. Figure 2 illustrates the

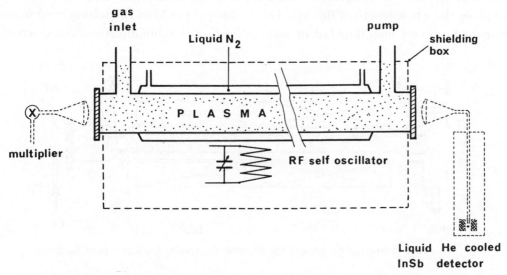

FIGURE 2. Schematic diagram of the radiofrequency (RF) discharge cell.

absorption cell, 1 m long with an internal diameter of 5 cm with cooling possible to liquid-nitrogen temperature. The plasma is excited by an RF oscillator delivering up to 1 kW of power at 50 MHz. About 50% of the power is coupled to the gas, and corresponds to a power density as high as 0.25 W cm^{-3}. Such high power is not necessary for the simple protonation of a stable molecule (Bogey *et al.* 1981) but is required to induce the relatively complex reactions that produce HCS$^+$ (Bogey *et al.* 1984*c*) or to populate highly excited vibrational states of neutrals (Bogey *et al.* 1982*a, b*, 1986*c, d*; Anacona *et al.* 1986) or ions (Bogey *et al.* 1983). Table 2 summarizes these results. Relative intensities of rotational transitions in successive vibrational states correspond to vibrational temperatures in the range 2000–4000 K, depending on the

TABLE 2. DIATOMIC MOLECULES OBSERVED BY MILLIMETRE-WAVE SPECTROSCOPY IN
VIBRATIONALLY EXCITED STATES

	v_{max}	E_v/cm^{-1}	$E_{dissociation}/cm^{-1}$	references
CO	40	66000	90000	Bogey *et al.* (1986*c*)
CS	20	23500	59000	Bogey *et al.* (1982*b*)
CN	9	18500	63000	Bogey *et al.* (1986*d*)
SO	8	9500	27000	Bogey *et al.* (1982*a*)
NS	5	6500	39000	Anacona *et al.* (1986)
CO$^+$	4	9500	67000	Bogey *et al.* (1983)

species of interest. From measurements of the plasma refractive index, the electron density has been estimated to be about 10^{10} cm^{-3}, which is of the same order of magnitude as in a positive column (Woods 1983).

3.2. *Spectroscopy in the negative glow*

As previously mentioned, the main attractive point of the negative glow, which develops near the cathode, is the high density of the plasma. Two discharges, the hollow cathode, and the magnetically confined glow, have been designed to lengthen the negative glow of a DC discharge.

In the hollow cathode, the primary electrons form an axial beam that enhances the ion density along the whole length of the cell. This technique has been extensively used in optical spectroscopy for a very long time but its first application to submillimetre-wave spectroscopy

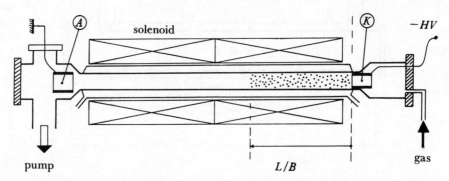

FIGURE 3. Schematic diagram of the magnetically confined negative glow discharge.

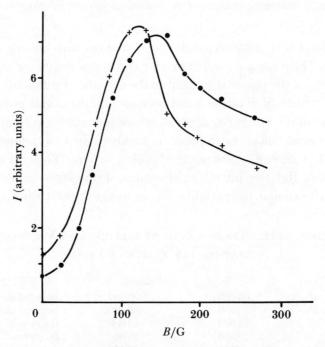

FIGURE 4. Characteristic variation of ionic line intensity against the confinement magnetic field. Symbols: ●, H_2D^+; +, $Ar \cdot H_3^+$. (1 G = 10^{-4} T.)

[78]

TABLE 3. MOLECULAR IONS OBSERVED BY TUNABLE MILLIMETRE- AND SUBMILLIMETRE-WAVE SPECTROSCOPY

ions	isotopes	methods of production	frequency range/GHz	experimental techniques	other methods	references to the first laboratory observation
CO^+	^{13}C, ^{18}O	PC, RF	118–1061	K, HG, SBL	IR, OPT	Dixon & Woods (1975)
HCO^+	D, ^{13}C, ^{18}O	PC, RF, HC	89–1070	K, HG, SBL	IR, IS	Woods et al. (1975)
HN_2^+	D, ^{15}N	PC, HC	90–1024	K, HG, SBL	IR, IS	Saykally et al. (1976a)
HCS^+	—	PC, RF	85–299	K, HG	IR, IS	Gudeman et al. (1981)
HOC^+	D, ^{18}O, ^{13}C	PC, MC	89–382	K, HG, BWO	IR	Gudeman & Woods (1982)
NO^+	—	MC	238–358	HG	OPT	Bowman et al. (1982)
ArH^+	D	MC	317–634	HG, SBL	IR	Bowman et al. (1983)
KrD^+	—	MC	252	HG	IR	Warner et al. (1984a)
H_2D^+	—	MC, HC	156–372	K, HG, BWO	IR, IS?	Bogey et al. (1984a); Warner et al. (1984b)
HCO_2^+	D	MC	340–405	BWO	IR, IS	Bogey et al. (1984b)
H_3O^+	—	MC	307–396	HG, BWO	IR, IS?	Bogey et al. (1985a); Plummer et al. (1985)
SO^+	—	DC, MC	69–250	K, HG	OPT	Warner et al. (1985)
$HCNH^+$	D	MC	148–370	HG, BWO	IR, IS	Bogey et al. (1985b)
OH^+	D	HC	909–1045	SBL	OPT, IR	Bekoy et al. (1985)
HON_2^+	D	MC	134–405	HG, BWO	IR	Bogey et al. (1986e)
NH^+	—	HC	1020	SBL	OPT	Verhoeve et al. (1986b)
CF^+	—	MC	102–410	HG	IR	Plummer et al. (1986)
HBF^+	D, ^{10}B	MC	145–363	HG	IR	Cazzoli et al. (1986)
ArH_3^+	D	MC	182–392	HG, BWO		Bogey et al. (1987)

Abbreviations: PC, positive column discharge; RF, radiofrequency discharge; HC, hollow-cathode discharge; MC, magnetically confined negative glow discharge; K, fundamental klystron; HG, harmonic generation; SBL, tunable side-band laser; BWO, backward wave oscillator; IR, infrared spectroscopy; OPT, optical spectroscopy; IS, interstellar detection.

is by Van den Heuvel & Dymanus (1982) who first observed rotational transitions of ions in the terahertz region. This device has been now used to study some light ions such as OH^+ (Bekooy et al. 1985), OD^+ (Verhoeve et al. 1986 a), NH^+ (Verhoeve et al. 1986 b), and ArH^+ (Laughlin et al. 1986). A similar device allowed Saito et al. (1985) to observe a millimetre-wave transition of H_2D^+.

In the magnetically lengthened negative glow, first developed by De Lucia & co-workers (1983), an axial magnetic field is used to confine the primary electrons and to limit ambipolar diffusion to the walls. The diagram, of the cell built in Lille is presented in figure 3. A magnetic field of about 200 G is generally sufficient to lengthen the negative glow to the 1 m length of our discharge cell. The discharge conditions correspond to those of the anormal glow discharge, i.e. high voltage across the discharge (2–5 kV) and low current (1–10 mA). Traces of reactant gases (10^{-4}–10^{-3} Torr†) are mixed with Ar used as a buffer gas, and the total pressure is generally lower than 10 mTorr. A variation of the intensity of an ion line with the confining magnetic field B is very characteristic: namely, a sharp initial increase with B, and the existence of an optimum magnetic field. This very typical behaviour is illustrated in figure 4, which shows the H_2D^+ line intensity dependence on the strength of the magnetic field. Such behaviour can be used to discriminate ion lines from neutral ones. The second curve in figure 4 definitely demonstrates that the line studied (unidentified at that time) could be ascribed to an ionic species, later identified as the ionic complex $Ar \cdot H_3^+$, as will be described in a following section.

This technique has proved to be very efficient, leading to the discovery of a variety of new ions in the last five years, from ions with very small dipole moments, NO^+ (Bowman et al. 1982), H_2D^+ (Warner et al. 1984 b; Bogey et al. 1984 a), to the first asymmetric tops, HCO_2^+ (Bogey et al. 1984 b, 1986 b), HON_2^+ (Bogey et al. 1986 e) or $Ar \cdot H_3^+$ (Bogey et al. 1987).

To conclude this section, table 3 summarizes the ions so far detected by tunable millimetre- and/or submillimetre-wave spectroscopy.

4. Spectroscopic studies of individual ionic species

The highly sensitive spectrometers and efficient methods of production described above have enabled us to detect many molecular ions, and we will consider in this section some examples that are characteristic of the specific problems encountered in these studies.

4.1. An ion with large amplitude motion: H_3O^+

This ion is one of the most fundamental molecular ions because of its important role in various physical–chemical processes reaching from biological systems (Eigen 1964) to interstellar clouds (Leung et al. 1984; Viala 1986). This ion is isoelectronic with NH_3 and has the same well-known umbrella shape, with a double minimum potential leading to an inversion splitting of the energy levels. Earliest theoretical calculations of the ground-state inversion splitting ranged from 28 to 72 cm^{-1} (Spirko & Bunker 1982; Botschwina et al. 1983; Bunker et al. 1983, 1984) making the prediction of the inversion rotation spectrum of this ion impossible. High-resolution IR spectra of H_3O^+ observed by different groups (Begeman et al. 1983; Haese & Oka 1984; Lemoine & Destombes 1984; Davies et al. 1985; Liu et al. 1985; Liu & Oka 1985) have

† 1 Torr $\approx$ 133.3 Pa.

indicated the frequency range where its inversion spectrum can be expected. Fortunately, the three $K = 0, 1, 2$ components of the $J = 3^+ \leftarrow 2^-$ were predicted to lie in the frequency range covered by our carcinotron and the search of these lines has been greatly facilitated by the good spectral purity and stability of the carcinotron used in the free running mode.

Three lines have been observed in the magnetically confined negative glow discharge in a mixture of H_2O (10^{-4} Torr), H_2 (4×10^{-4} Torr), and Ar (15×10^{-3} Torr) at room temperature, and have exhibited the typical intensity variation against the confining magnetic field expected for ion lines. This is well illustrated in figure 5. The three frequencies were well within the

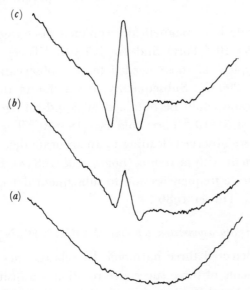

FIGURE 5. Evolution of the $J, K = 3^+, 0 \leftarrow 2^-, 0$ transition of H_3O^+ against magnetic field (396 272.412 MHz). The discharge current has been kept constant at $I = 4$ mA; $p_{H_2O} = 10^{-4}$ Torr; $p_{H_2} = 4 \times 10^{-4}$ Torr; $p_{Ar} = 15 \times 10^{-3}$ Torr. Lock-in time constant $= 100$ ms. The width of the region shown is about 8 MHz. (a) 0 G; (b) 75 G; (c) 150 G.

uncertainties given by Liu & Oka (1985) and their assignment to H_3O^+ lines has been confirmed by chemical evidence (Bogey et al. 1985 a). At the same time the $J, K = 1^-, 1 \leftarrow 2^+$, 1 line was observed by Plummer et al. (1985) in similar experimental conditions. These four lines, although accurately measured, permitted only the determination of a linear combination of various molecular parameters. Then, a global fit of both submillimetre and IR data led to a significant improvement of the molecular constant determination (Liu et al. 1986).

The $J, K = 1^-, 1 \leftarrow 2^+$, 1 transition, which connects relatively low-lying energy levels, has been tentatively detected in two astronomical sources (Wootten et al. 1986; Hollis et al. 1986). The observation of the $J = 3^+ \leftarrow 2^-$ components would confirm this identification and would place H_3O^+ among the most abundant molecular ions in the interstellar gas, in agreement with theoretical predictions (Leung et al. 1984; Viala 1986).

4.2. A weakly polar ion: HCNH⁺

Although it is much less stable than HCN in laboratory conditions, the isomer HNC has been detected in several interstellar sources with an abundance comparable to that of HCN. This is strong support for a production mechanism involving dissociative recombination of the protonated molecule $HCNH^+$ (Leung *et al.* 1984). In view of its importance in astrochemistry, this ion has been the subject of numerous *ab initio* calculations to provide accurate rotational frequencies (Allen *et al.* 1980; Haese & Woods 1979; Dardi & Dykstra 1980). However, because of the small calculated dipole moment (0.3 D or less) (Haese & Woods 1979; Dardi & Dykstra 1980; Lee & Schaefer 1984), its detection by millimetre-wave spectroscopy was expected to be difficult. Although ions generally have a rather large dipole moment, some weakly polar ions have also been observed (NO^+, Bowman *et al.* 1982; H_2D^+, Bogey *et al.* 1984*a*; Warner *et al.* 1984*b*; Saito *et al.* 1985). For $HCNH^+$, IR detection (Altman *et al.* 1984*a*, *b*; Amano 1984) has provided accurate rotational parameters that facilitated the submillimetre search.

First observations were made in a magnetically confined discharge through a mixture of HCN (8×10^{-4} Torr), H_2 (6×10^{-3} Torr) and Ar (15×10^{-3} Torr) cooled to 220 K. The discharge conditions were typical of those needed for the observation of ions ($I = 2$ mA; $HV = 2200$ V; optimum $B = 200$ G). Subsequently, it was found that an improvement of about 5 in the signal:noise ratio could be obtained at liquid-nitrogen temperature by discharging a mixture of CH_4 (1.5×10^{-3} Torr) and N_2 (18×10^{-3} Torr). By using harmonic generation, two more lines were observed, leading to an accurate determination of the B and D values in very good agreement with IR results (Bogey *et al.* 1985*b*). Both IR and millimetre-wave data finally led to accurate frequencies and to subsequent detection of $HCNH^+$ in the interstellar medium (Ziurys & Turner 1986).

4.3. *Slightly asymmetric-top ions*: $HOCO^+$, HON_2^+

In 1981, Thaddeus *et al.* detected three harmonically related lines in Sagr. B2 that were interpreted as $K = 0$ components of three successive rotational transitions of a nearly linear molecule. By comparing the deduced B value with *ab initio* calculations for a number of plausible molecules, they concluded that the most likely candidates for these lines were isoelectronic molecules, the protonated carbon dioxide $HOCO^+$ and the cyanic acid HOCN. The most conclusive way to solve this problem was to observe one or the other species in the laboratory. In a first step, we observed six lines in the 350–380 GHz range by discharging a mixture of CO_2 (2×10^{-5} Torr), H_2 (10^{-4} Torr) and Ar (6×10^{-3} Torr) slightly cooled by flowing cold N_2 gas around the cell. These lines were identified as the $K = 0$ and $K = 2$ components of the $J = 16 \leftarrow 15$ and $J = 17 \leftarrow 16$ transitions of $HOCO^+$. They fitted very nicely with the three astrophysical lines, giving a definite identification of the astrophysical carrier.

With the phase-locked carcinotron, a more extensive study of the spectrum has been carried out and higher K components have been observed in the same frequency range. By replacing H_2 by D_2 in the discharge, we have also been able to measure a number of transitions of $DOCO^+$ (Bogey *et al.* 1986*b*). At the same time, the first IR spectra were obtained by Amano & Tanaka (1985) with a hollow cathode discharge.

Protonated carbon dioxide is a very slightly asymmetric prolate rotor ($\kappa = -0.9996$) and because of the large value of the rotational constant A, only two different types of transitions can be observed in the frequency range of our submillimetre-wave source: a-type R-branch and b-type P- or R-branch transitions. Currently, only a-type ${}^qR_{K_a}$ transitions have been observed. They are characterized by $15 \leqslant J \leqslant 18$, $K_a \leqslant 5$ for $HOCO^+$ (29 lines) and $16 \leqslant J \leqslant 19$, $K_a \leqslant 6$ for $DOCO^+$ (31 lines). They have been fitted by using extended Watson's S-reduced hamiltonian in the I^r-axis representation, and a set of accurate molecular parameters has been determined that agrees with the predictions of *ab initio* calculations

(De Frees *et al.* 1982; Frisch *et al.* 1985) (table 4). A comparison of theoretical and experimental values permits the improvement of the prediction of the molecular constants of the ^{13}C isotopic species, which will be searched for in our laboratory in the near future.

According to *ab initio* calculations, the isoelectronic molecule HON_2^+ has a similar structure, i.e. a nearly linear backbone with the hydrogen slightly off axis, and the preferred site of protonation is the oxygen atom. However, the nitrogen-protonated form is predicted to be also stable and about 11.5 kcal mol^{-1}† above the oxygen-protonated form (Rice *et al.* 1986; Yamashita & Morokuma 1986; Vincent & Hillier 1986). Recently, Amano (1986) has measured the ν_1 IR band and has deduced molecular constants in agreement with those of the O-protonated form, permitting us to search for its rotational spectrum in the submillimetre-wave range.

Relatively strong lines belonging to a-type ^qR- and b-type ^rR-, ^qR- and ^rP-branches were observed in a mixture of N_2O (2×10^{-3} Torr), H_2 (7×10^{-3} Torr) and Ar (9×10^{-3} Torr). The optimum discharge conditions were $I = 5$ mA, $HV = 1600$ V, $B = 250$ G (Bogey *et al.* 1986*e*). By replacing H_2 with D_2, the DON_2^+ lines were easily found by using the carcinotron in the free running mode. Figure 6 illustrates the main advantage of this submillimetre source that enabled us to have an overview of the spectrum with a high sensitivity permitting the straightforward identification of the lines.

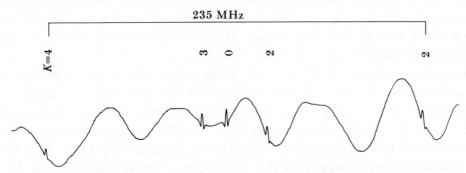

FIGURE 6. Overview of the DON_2^+ spectrum around 275 GHz showing the different K components of the $J = 13 \leftarrow 12$ transition. The carcinotron is used in the free-running mode.

TABLE 4. SPECTROSCOPIC CONSTANTS FOR THE GROUND STATE OF HCO_2^+ AND DCO_2^+ FROM EXTENDED WATSON'S S-REDUCED HAMILTONIAN IN THE I_r AXIS REPRESENTATION

A/MHz	788629 (3070)[a]	433038 (178)
B/MHz	10733.7221 (89)	10163.9589 (91)
C/MHz	10609.4435 (87)	9908.6414 (92)
D_J/kHz	3.5029 (33)	3.0999 (36)
D_{JK}/kHz	935.0 (18)	319.87 (12)
d_1/kHz	−0.0426 (70)	−0.1294 (63)
d_2/kHz	−0.0145 (22)	−0.0738 (67)
H_{KJ}/Hz	3786. (314)	−364.4 (40)
h_2/Hz	0[b]	0.057 (12)
L_{KJ}/Hz	79. (19)	0[b]
S_{KJ}/Hz	1.79 (38)	0[b]
κ	−0.9996	−0.9988

[a] Numbers in parentheses denote one standard deviation in the last digits.
[b] These constants were poorly determined and were therefore set equal to zero.

† 1 kcal = 4.184×10^3 J.

A total number of 63 rotational lines for HON_2^+ and 89 for DON_2^+ characterized respectively by $5 \leqslant J \leqslant 18$, $K_a \leqslant 4$ and $0 \leqslant J \leqslant 17$, $K_a \leqslant 7$, were measured in the 130–405 GHz frequency range. They have been fitted by using Watson's S-reduced hamiltonian, and the spectra of both HON_2^+ and DON_2^+ are reproduced within 23 kHz. Because both a-type and b-type transitions have been measured, an accurate set of molecular parameters has been determined. This set is as complete as is available for the isoelectronic stable molecule HNCO (Hocking *et al.* 1975). The results will be published elsewhere.

Finally, note that an extensive search for the N-protonated form has been unsuccessful.

4.4. *A weakly bound ionic cluster:* $Ar \cdot H_3^+$

Gas-phase ionic clusters are present in a number of physical-chemical systems, such as in flames or in the upper atmosphere, and are generally though to play an important role in situation where ions are involved and where stabilization of the complex can occur via collisions or radiative transitions (Castleman & Keesee 1986). Although there is a growing interest in the study of the dynamics of their formation and photodissociation, very little is known about the spectroscopic properties of these species. In addition to the pioneering work of Schwarz (1977), only a few low-resolution spectra have been obtained: in the visible, for clusters like $Ar \cdot C_6F_6^+$ by Dimauro *et al.* (1984) and in the infrared, for clusters like H_{2n+1}^+ or $H_3O^+(H_2O)_n$ by Okumura *et al.* (1985). In particular, information on the structure of such complexes comes essentially from *ab initio* calculations (Raynor & Herschbach 1983; Ahlrichs 1975; Yamagucchi *et al.* 1983; Frisch *et al.* 1986).

In course of a search for CH_2D^+, a weakly polar variant of CH_3^+ that is thought to play a key role in interstellar chemistry, we detected a line near 390 GHz that was ascribed to an ionic species in view of the dependence of its intensity against the confinement magnetic field (see figure 4). However, it was immediately recognized that CH_4 was not at all necessary to produce this ion. Moreover, most of the gases we added to the discharge (N_2, O_2, C_2H_2, CO, Kr) led to the immediate disappearance of the line, even when added in trace amounts. In contrast, addition of H_2 and of compounds with a proton affinity lower than that of hydrogen (Ne, He) led only to a slight decrease of the line intensity. Clearly the ion was formed with only Ar and/ or D atoms. By scanning ± 2.5 GHz around the line frequency we discovered only four additional lines. This set of five lines showed the characteristic K pattern expected for a $J \geqslant 2$ a-type ^qR rotational transition of a slightly asymmetric prolate rotor. By carefully exploring a frequency range of about 40 GHz with the free running carcinotron, we found another characteristic set of five lines. A fit of these 10 lines with Watson's A-reduced hamiltonian led to the determination of the molecular constants A, B, C and of four centrifugal distortion parameters. The value of the inertial defect ($\Delta = 0.11$ u Å^2†) clearly indicated that the molecule was planar and the relative intensities of the K components were consistent with a C_{2v} symmetry. Comparison of the A value (655 137 MHz) with that of D_3^+ (654 266 MHz) strongly suggested that a D_3^+ ion was involved in the molecule; this was also supported by the similar intensity behaviour of the H_2D^+ line and of the unknown lines against the confinement magnetic field (figure 4). The relatively low values of B and C, however, suggested that one heavy atom was involved in the molecule, most probably Ar in view of previously mentioned chemical evidence.

† 1 Å $= 10^{-10}$ m $= 10^{-1}$ nm.

By analogy with the $He \cdot H_3^+$ and $H_2 \cdot H_3^+$ clusters, well characterized by *ab initio* calculations (Poshusta & Agrawal 1973; Ahlrichs 1975; Raynor & Herschbach 1983; Dykstra 1983; Yamaguchi *et al.* 1983), we assumed that the Ar atom lay on a symmetry axis of the D_3^+ triangle, which we assumed to be little disturbed by the presence of Ar. By adjusting the distance between Ar and D_3^+, we were able to account for the molecular constants within 3×10^{-3} and to predict the $Ar \cdot H_3^+$ spectrum, which was sought by just replacing D_2 by H_2. The $Ar \cdot H_3^+$ lines were found very close to the predicted frequencies (within less than 2×10^{-3}), thus confirming the identification of the ion and of its structure (Bogey *et al.* 1987).

$Ar \cdot H_3^+$ appears to be planar with the Ar lying on a symmetry axis of H_3^+ at 2.38 Å from the H_3^+ centroid. From such a structure, a very large dipole moment is expected (of the order of 9 D) and explains why we detected this complex in spite of experimental conditions *a priori* not favourable to cluster production.

Study of the mixed isotopic forms $Ar \cdot H_2D^+$ and $Ar \cdot HD_2^+$ is now in progress and some lines have already been observed at frequencies in complete agreement with those predicted with the structure given above. A number of the lines observed for the four isotopic forms are in fact doublets (figure 7), showing that tunnelling motions occur. When quantitatively interpreted, these results will lead to the determination of an accurate substitution structure and will shed light on the dynamics of such ionic clusters, a first step towards the study of other complexes such as $H_2 \cdot H_3^+$.

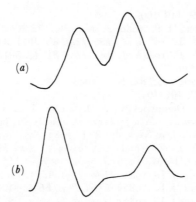

(a)

(b)

FIGURE 7. Two examples of tunnelling motion doublets observed in the $Ar \cdot H_3^+$ complex. *(a)* The $12_{0,12} \leftarrow 11_{0,11}$ transition of $Ar \cdot D_3^+$ near 390.3 GHz. The splitting is 800 kHz. *(b)* The $9_{0,9} \leftarrow 8_{0,8}$ transition of $Ar \cdot HDH^+$ near 399.4 GHz. The splitting is 5.5 MHz.

5. CONCLUSION

Major improvements in ion-production methods as well as in spectrometer sensitivity and frequency coverage have made possible the recent impressive developments in the field of ion high-resolution rotational spectroscopy. In particular, the close coupling between radioastronomy and laboratory spectroscopy, which have been intimately linked since the detection of the first ion, have led in the last three years to the interstellar detection of fundamental species such as H_2D^+ (Phillips *et al.* 1985), $HCNH^+$ (Ziurys *et al.* 1986) and H_3O^+ (Hollis *et al.* 1986; Wootten *et al.* 1986), thus giving new insights into interstellar chemistry. Spectroscopic studies of molecular ions can now be as complete as those of stable molecules, and molecular structures of ions have been derived (Woods *et al.* 1981; Bogey *et al.* 1981; Szanto *et al.* 1981; Gudeman

& Woods 1982; Bogey *et al.* 1986*a*). Further developments can now be expected from the interactions between the very active fields of radioastronomy, *ab initio* calculations, infrared (IR) and millimetre-wave spectroscopy. The detection of new objects, which was anticipated for a few years (Woods 1983; Gudeman & Saykally 1984) is now a reality, because negative ions, which have already been detected in IR (Owrutsky *et al.* 1985; Liu & Oka 1986; Tack *et al.* 1986; Kawaguchi & Hirota 1986; Brown *et al.* 1986; Gruebele *et al.* 1987) will, without doubt, be observed in the millimetre-wave range in the near future. The recent detection of the $Ar \cdot H_3^+$ ionic cluster in a plasma is a stimulating challenge for both experimentalists and theoreticians to detect other similar species.

We gratefully acknowledge the financial support of the Centre National de la Recherche Scientifique and of the Établissement Public Régional Nord-Pas de Calais. The submillimetre-wave backward oscillator was lent by the Centre Commun de Mesures de l'Université de Lille Flandres Artois, and the Schottky barrier diodes have been provided by the DEMIRM (Observatoire de Paris-Meudon).

Many thanks are also due to H. Bolvin, M. Denis and B. Lemoine for their participation in this work, and to P. Rosseels for technical assistance.

REFERENCES

Ahlrichs, R. 1975 *Theor. chim. Acta* **39**, 149–160.
Allen, T. L., Goddard, J. D. & Schaefer, H. F. 1980 *J. chem. Phys.* **73**, 3255–3263.
Altman, R. S., Crofton, M. W. & Oka, T. 1984*a* *J. chem. Phys.* **80**, 3911–3912.
Altman, R. S., Crofton, M. W. & Oka, T. 1984*b* *J. chem. Phys.* **81**, 4255–4258.
Amano, T. 1984 *J. chem. Phys.* **81**, 3350–3351.
Amano, T. & Tanaka, K. 1985 *J. chem. Phys.* **82**, 1045–1046.
Amano, T. 1986 *Chem. Phys. Lett.* **127**, 101–104.
Anacona, J., Bogey, M., Davies, P. B., Demuynck, C. & Destombes, J. L. 1986 *Molec. Phys.* **59**, 81–88.
Badareu, E. & Popescu, I. 1968 *Gaz ionisés – décharges électriques dans les gaz*, pp. 91–147. France: Éditions Dunod.
Begeman, M. H., Gudeman, C. S., Pfaff, J. & Saykally, R. J. 1983 *Phys. Rev. Lett.* **51**, 554–557.
Bekooy, J. P., Verhoeve, P., Meerts, W. L. & Dymanus, A. 1985 *J. chem. Phys.* **82**, 3868–3869.
Blake, G. A., Farhoomand, J. & Pickett, H. M. 1986 *J. molec. Spectrosc.* **115**, 226–228.
Bogey, M., Demuynck, C. & Destombes, J. L. 1981 *Molec. Phys.* **43**, 1043–1050.
Bogey, M., Demuynck, C. & Destombes, J. L. 1982*a* *Chem. Phys.* **66**, 99–104.
Bogey, M., Demuynck, C. & Destombes, J. L. 1982*b* *J. molec. Spectrosc.* **95**, 35–42.
Bogey, M., Demuynck, C. & Destombes, J. L. 1983 *J. chem. Phys.* **79**, 4704–4707.
Bogey, M., Demuynck, C., Denis, M., Destombes, J. L. & Lemoine, B 1984*a* *Astron. Astrophys.* **137**, L15–L16.
Bogey, M., Demuynck, C. & Destombes, J. L. 1984*b* *Astron. Astrophys.* **138**, L11–L12.
Bogey, M., Demuynck, C. & Destombes, J. L. 1984*c* *J. molec Spectrosc.* **107**, 417–418.
Bogey, M., Demuynck, C., Denis, M. & Destombes, J. L. 1985*a* *Astron. Astrophys.* **148**, L11–L13.
Bogey, M., Demuynck, C. & Destombes, J. L. 1985*b* *J. chem. Phys.* **83**, 3703–3705.
Bogey, M., Demuynck, C. & Destombes, J. L. 1986*a* *J. molec. Spectrosc* **115**, 229–231.
Bogey, M., Demuynck, C. & Destombes, J. L. 1986*b* *J. chem. Phys.* **84**, 10–15.
Bogey, M., Demuynck, C., Destombes, J. L. & Lapauw, J. M. 1986*c* *J. Phys.* **E19**, 520–525.
Bogey, M., Demuynck, C. & Destombes, J. L. 1986*d* *Chem. Phys.* **102**, 141–146.
Bogey, M., Demuynck, C., Destombes, J. L. & McKellar, A. R. W. 1986*e* *Astron. Astrophys.* **167**, L13–L14.
Bogey, M., Demuynck, C. & Destombes, J. L. 1986*f* *Chem. Phys. Lett.* **125**, 383–388.
Bogey, M., Bolvin, H., Demuynck, C. & Destombes, J. L. 1987 *Phys. Rev. Lett.* **58**, 988–991.
Botschwina, P., Rosmus, P. & Reinsch, E. A. 1983 *Chem. Phys. Lett.* **102**, 299–306.
Bowman, W. C., Herbst, E. & De Lucia, F. C. 1982 *J. chem. Phys.* **77**, 4261–4262.
Bowman, W. C., Plummer, G. M., Herbst, E. & De Lucia, F. C. 1983 *J. chem. Phys.* **79**, 2093–2095.
Brown, R. D., Godfrey, P. D., McGilbery, D. C. & Croeft, J. G. 1981 *Chem. Phys. Lett.* **84**, 437–439.
Brown, P. R., Davies, P. B. & Johnson, S. A. 1986 *Chem. Phys. Lett.* **132**, 582–584.
Buhl, D. & Snyder, L. E. 1970 *Nature, Lond.* **228**, 267–269.
Bunker, P. R., Kraemer, W. P. & Spirko, V. 1983 *J. molec. Spectrosc.* **101**, 180–185.

Bunker, P. R., Amano, T. & Spirko, V. 1984 *J. molec. Spectrosc.* **107**, 208–211.

Bustreel, R., Demuynck-Marliere, C., Destombes, J. L. & Journel, G. 1979 *Chem. Phys. Lett.* **67**, 178–182.

Castleman, A. W. & Keesee, R. G. 1986 *Chem. Rev.* **86**, 589–618.

Cazzoli, G., Degli Esposti, C., Dore, L. & Favero, P. G. 1986 *J. molec. Spectrosc.* **119**, 467.

Dardi, P. S. & Dykstra, C. E. 1980 *Astrophys. J.* **240**, L171–L173.

Davies, P. B., Hamilton, P. A. & Johnson, S. A. 1985 *J. opt. Soc. Am.* **2**, 794–799.

De Frees, D. J., Loew, G. H. & McLean, A. D. 1982 *Astrophys. J.* **254**, 405–411.

De Lucia, F. C., Herbst, E., Plummer, G. M. & Blake, G. A. 1983 *J. chem. Phys.* **78**, 2312–2316.

Dimauro, L. F., Heaven, M. & Miller, T. A. 1984 *Chem. Phys. Lett.* **104**, 526–532.

Dixon, T. A. & Woods, R. C. 1975 *Phys. Rev. Lett.* **34**, 61–63.

Dixon, T. A. & Woods, R. C. 1977 *J. chem. Phys.* **67**, 3956–3964.

Dykstra, C. E. 1983 *J. molec. Struct.* **103**, 131–138.

Eigen, M. 1964 *Angew. Chem. Int. Ed. Engl.* **3**, 1–72.

Evenson, K. M., Jennings, D. A. & Petersen, F. R. 1984 *Appl. Phys. Lett.* **44**, 576–578.

Frisch, M. J., Schaefer, H. F. & Binkley, J. S. 1985 *J. phys. Chem.* **89**, 2192–2194.

Frisch, M. J., Del Bene, J. E., Binkley, J. S. & Schaeffer, H. F. 1986 *J. chem. Phys.* **84**, 2279–2289.

Green, S., Montgomery, J. A. Jr & Thaddeus, P. 1974 *Astrophys. J.* **193**, L89–L92.

Gruebele, M., Polak, M. & Saykally, R. J. 1987 *J. chem. Phys.* **86**, 1698–1702.

Gudeman, C. S., Haese, N. N., Piltch, N. D. & Woods, R. C. 1981 *Astrophys. J.* **246**, L47–L49.

Gudeman, C. S. 1982 Ph.D. thesis, University of Wisconsin.

Gudeman, C. S. & Woods, R. C. 1982 *Phys. Rev. Lett.* **48**, 1344–1348.

Gudeman, C. S. & Saykally, R. J. 1984 *A. Rev. phys. Chem.* **35**, 387–418.

Haese, N. N. & Woods, R. C. 1979 *Chem. Phys. Lett.* **61**, 396–398.

Haese, N. N. & Oka, T. 1984 *J. chem. Phys.* **80**, 572–573.

Hirota, E. 1985 *High resolution spectroscopy of transient molecules.* Berlin: Springer Verlag.

Hocking, W. H., Gerry, M. C. L. & Winnewisser, G. 1975 *Can. J. Phys.* **53**, 1869–1901.

Hollis, J. M., Churchwell, E. B., Herbst, E. & De Lucia, F. C. 1986 *Nature, Lond.* **322**, 524–526.

Johnson, M. A., Alexander, M. L., Hertel, I. & Lineberger, W. C. 1984 *Chem. Phys. Lett.* **105**, 374–379.

Kawaguchi, K. & Hirota, E. 1986 *J. chem. Phys.* **84**, 2953–2960.

Klemperer, W. 1970 *Nature, Lond.* **227**, 1230.

Kraemers, W. P. & Diercksen, G. H. F. 1976 *Astrophys. J.* **205**, L97–L100.

Laughlin, K. B., Blake, G. A., Cohen, R. & Saykally, R. J. 1986 In *41st Annual Symposium on Molecular Spectroscopy, Columbus, Ohio*, paper FA2. Ohio State University.

Lee, T. J. & Schaeffer, H. F. 1984 *J. chem. Phys.* **80**, 2977–2978.

Lemoine, B. & Destombes, J. L. 1984 *Chem. Phys. Lett.* **111**, 284–287.

Leung, C. M., Herbst, E. & Huebner, W. F. 1984 *Astrophys. J. Suppl. Ser.* **56**, 231–256.

Liu, D. J. & Oka, T. 1985 *Phys. Rev. Lett.* **54**, 1787–1789.

Liu, D. J., Haese, N. N. & Oka, T. 1985 *J. chem. Phys.* **82**, 5368–5372.

Liu, D. J. & Oka, T. 1986 *J. chem. Phys.* **84**, 2426–2427.

Liu, D. J., Oka, T. & Sears, T. J. 1986 *J. chem. Phys.* **84**, 1312–1316.

Low, W. & Ramberg, Y. 1955 *Bull. Res. Coun. Israel* **5A**, 40–45.

Okumura, M., Yeh, L. I. & Lee, Y. T. 1985 *J. chem. Phys.* **83**, 3705–3706.

Owrutsky, J. C., Rosenbaum, N. H., Tack, L. M. & Saykally, R. J. 1985 *J. chem. Phys.* **83**, 5338–5339.

Phillips, T. G., Blake, G. A., Keene, J., Woods, R. C. & Churchwell, E. 1985 *Astrophys. J.* **294**, L45–L48.

Plummer, G. M., Herbst, E. & De Lucia, F. C. 1985 *J. chem. Phys.* **83**, 1428–1429.

Plummer, G. M., Anderson, T., Herbst, E. & De Lucia, F. C. 1986 *J. chem. Phys.* **84**, 2427–2428.

Poshusta, R. D. & Agrawal, V. P. 1973 *J. chem. Phys.* **59**, 2477–2482.

Raynor, S. & Herschbach, D. R. 1983 *J. phys. Chem.* **87**, 289–293.

Rice, J. E., Lee, T. J. & Schaefer, H. F. 1986 *Chem. Phys. Lett.* **130**, 333–336.

Saito, S., Kawaguchi, K. & Hirota, E. 1985 *J. chem. Phys.* **82**, 45–47.

Sastry, K. V. L. N., Herbst, E. & De Lucia, F. C. 1981 *J. chem. Phys.* **75**, 4169–4170.

Saykally, R. J., Dixon, T. A., Anderson, T. G., Szanto, P. G. & Woods, R. C. 1976a *Astrophys. J.* **205**, L101–L103.

Saykally, R. J., Szanto, P. G., Anderson, T. G. & Woods, R. C. 1976b *Astrophys. J.* **204**, L143–L145.

Saykally, R. J. 1977 Ph.D. thesis, University of Wisconsin.

Schwartz, H. A. 1977 *J. chem. Phys.* **67**, 5525–5534.

Skatrud, D. D., De Lucia, F. C., Blake, G. A. & Sastry, K. V. L. N. 1983 *J. molec. Spectrosc.* **99**, 35–46.

Snyder, L. E., Hollis, J. M., Lovas, F. J. & Ulich, B. L. 1976 *Astrophys. J.* **209**, 67–74.

Spirko, V. & Bunker, P. R. 1982 *J. molec. Spectrosc.* **95**, 226–235.

Szanto, P. G., Anderson, T. G., Saykally, R. J., Piltch, N. D., Dixon, T. A. & Woods, R. C. 1981 *J. chem. Phys.* **75**, 4261–4263.

Tack, L. M., Rosenbaum, N. H., Owrutski, J. C. & Saykally, R. J. 1986 *J. chem. Phys.* **84**, 7056–7057.

Tang, T. B., Inokuchi, H., Saito, S., Yamada, C. & Hirota, E. 1985 *Chem. Phys. Lett.* **116**, 83–85.

Thaddeus, P., Guelin, M. & Linke, R. A. 1981 *Astrophys. J.* **246**, L41–L45.

Thaddeus, P., Vrtilek, J. M. & Gottlieb, C. A. 1986 *Astrophys. J.* **299**, L63–L66.

Turner, B. E. 1974 *Astrophys. J.* **193**, L83–L86.

Van Den Heuvel, J. C., Meerts, W. L. & Dymanus, A. 1980 *J. molec. Spectrosc.* **84**, 162–169.

Van Den Heuvel, J. C. & Dymanus, A. 1982 *Chem. Phys. Lett.* **92**, 219–222.

Verhoeve, P., Bekooy, J. P., Meerts, W. L., Ter Meulen, J. J. & Dymanus, A. 1986a *Chem. Phys. Lett.* **125**, 286–289.

Verhoeve, P., Ter Meulen, J. J., Meerts, W. L. & Dymanus, A. 1986b *Chem. Phys. Lett.* **132**, 213–217.

Verma, U. P., Möller, K., Vogt, J., Winnewisser, M. & Christiansen, J. J. 1985 *Can J. Phys.* **63**, 1173–1183.

Viala, Y. P. 1986 *Astron. Astrophys. Suppl.* **64**, 391–437.

Vincent, M. & Hillier, I. H. 1986 *Chem. Phys. Lett.* **130**, 330–336.

Von Engel, A. 1965 *Ionized gases*, pp. 217–258. Oxford: Clarendon Press.

Wahlgreen, U., Liu, B., Pearson, P. K. & Schaefer, H. F. 1973 *Nature, Lond.* **246**, 4–6.

Warner, H. E., Conner, W. T. & Woods, R. C. 1984a *J. chem. Phys.* **81**, 5413–5416.

Warner, H. E., Conner, W. T., Petrmichl, R. H. & Woods, R. C. 1984b *J. chem. Phys.* **81**, 2514.

Warner, H. E., Carballo, N. & Woods, R. C. 1985 In *40th Annual Symposium on Molecular Spectroscopy, Columbus, Ohio*. Ohio State University.

Woods, R. C. & Dixon, T. A. 1974 *Rev. scient. Instrum.* **45**, 1122–1126.

Woods, R. C., Dixon, T. A., Saykally, R. J. & Szanto, P. G. 1975 *Phys. Rev. Lett.* **35**, 1269–1271.

Woods, R. C., Saykally, R. J., Anderson, T. G., Dixon, T. A. & Szanto, P. G. 1981 *J. chem. Phys.* **75**, 4256–4260.

Woods, R. C. 1983 In *Molecular ions: spectroscopy, structure and chemistry* (ed. T. A. Miller & V. E. Bondybey), pp. 11–47. Amsterdam: North Holland.

Wootten, A., Boulanger, F., Bogey, M., Combes, F., Encrenaz, P. J., Gerin, M. & Ziurys, L. 1986 *Astron. Astrophys.* **166**, L15–L18.

Yamaguchi, Y., Gaw, J. F. & Schaefer, H. F. 1983 *J. chem. Phys.* **78**, 4074–4085.

Yamashita, K. & Morokuma, K. 1986 *Chem. Phys. Lett.* **131**, 237–242.

Ziurys, L. M. & Turner, B. E. 1986 *Astrophys. J.* **302**, L31–L36.

Discussion

E. Hirota (*Institute for Molecular Science, Okazaki* 444, *Japan*). Professor Destombes has explained the doublet structure observed for some lines of $Ar \cdot D_3^+$ in terms of the internal rotation of the D_3 group. Then he should observe the effect of the spin statistics on the doublets. Did he really observe such statistical weights?

J. L. Destombes. The cnpi group of $Ar \cdot D_3^+$ is G_{12} and it can be used to predict the statistical weights of the energy levels. The observed line intensity ratios of the doublets are in good agreement with these predictions (see table IV in Bogey *et al.* 1987). The same agreement has been found for the other isotopic forms.

Phil. Trans. R. Soc. Lond. A **324**, 163–178 (1988)

Printed in Great Britain

High-resolution infrared spectroscopy of molecular ions

By T. Amano

Herzberg Institute of Astrophysics, National Research Council, Ottawa, Ontario, Canada K1A 0R6

Experimental techniques of high-resolution infrared absorption spectroscopy are described with an emphasis on the modulated hollow-cathode discharge techniques. Among the recent results, the ν_1 fundamental band of $HOCS^+$ ($\nu_0 = 3435.16$ cm^{-1}) and the $X^2\Pi$ and $a^4\Sigma^-$ system of NH^+ ($\nu_0 = 2903.17$ cm^{-1} in $X^2\Pi$ state, $\nu_0 = 2544.30$ cm^{-1} in $a^4\Sigma^-$ state, and $\Delta E(a^4\Sigma^- - X^2\Pi) = 323.90$ cm^{-1}) are presented to illustrate the quality and type of information that the high-resolution ion spectroscopy can provide. Another topic discussed is the abundance ratio $[HCO^+]/[HOC^+]$ as measured in the laboratory and its astrophysical implications. The measurement has been made possible by our recent detection of the ν_1 fundamental band of HOC^+ ($\nu_0 = 3268.03$ cm^{-1}). A rate-equation analysis was performed by assuming a reaction scheme to derive the rate constants of several key reactions.

1. Introduction

High-resolution infrared spectroscopy of molecular ions has been one of the most dynamic fields in molecular spectroscopy since the detection of the ν_2 fundamental band of H_3^+ by Oka (1980). Almost all recent observations of ions in the gas phase have been achieved by using tunable infrared laser sources such as difference-frequency lasers, colour-centre lasers, and diode lasers. A breakthrough was brought about by the advent of the velocity-modulation technique (Gudeman *et al.* 1983) and the modulated hollow-cathode discharge (Foster *et al.* 1984). Without these techniques, progress would have been very much slower, because ionic species are of very small concentration and are buried among much more stable and more abundant neutral species. The quality of data obtained with those tunable laser sources is very high and the amount of information is plentiful, because of their high resolution and precision. Rotational constants determined by the infrared analyses are often accurate enough to predict the rotational transition frequencies within a few megahertz to assist microwave spectroscopists to search for the lines in the laboratory.

A few review papers appeared in the past several years. Gudeman & Saykally listed 14 ions including HD^+, CH^+, and KrH^+ in their review article (1984). Kawaguchi & Hirota's review paper (1985) included 22 ions with some deuterated species counted separately. A more recent review by Sears (1987) cited 41 ions (excluding HD^+, CH^+, and KrH^+). Now 12 more positive and negative ions and several more new bands are to be added to Sears's compilation as of February, 1987. Table 1 is a supplement to the table provided by Sears. This growing list of the ions studied with tunable infrared laser techniques is an indication of how rapid the progress has been.

It is not possible to discuss here all the results obtained so far, even though they are limited to the newest ones in the past year. In this article, experimental techniques of difference-frequency laser spectroscopy with a hollow-cathode discharge as an ion source and several new results obtained with this technique are presented.

TABLE 1. MOLECULAR IONS STUDIED BY TUNABLE INFRARED LASER SPECTROSCOPY

(Additional to the table compiled by Sears (1987).)

diatomics

molecule	comments[a]	references
NH^+	(1–0) bands in a and X states (DF)	Kawaguchi & Amano (1987a)
SH^+	(1–0) band (DL)	Brown et al. (1986)
SH^-	(1–0) band (DL)	Greubele et al. (1987a)

triatomics

molecule	vibrational mode	comments	references
HOC^+	ν_1	DF	Nakanaga & Amano (1987b)
	$\nu_1+\nu_2-\nu_2$		Amano (1987)
HBF^+	ν_1; ν_3	DF; DL	Kawaguchi & Amano (1987b); Kawaguchi & Hirota (1986a)
H_2Cl^+	ν_1, ν_3; ν_2	DF; DL	Lee et al. (1987); Kawaguchi & Hirota (1986b)
NH_2^+	ν_3	DF	Rehfuss et al., personal communication[b]
N_3^-	ν_3	DL	Polak et al. (1987)
NCO^-	ν_3	DL	Greubele et al. (1987b)
NCS^-	ν_1	DL	Greubele et al., personal communication[c]

tetra-atomics

molecule	vibrational mode	comments	references
$HOCS^+$	ν_1	DF	Nakanaga & Amano (1987c)
$C_2H_2^+$	ν_3	DF	Crofton et al. (1987)
$HCNH^+$	ν_1, ν_2; ν_3	DF	Attman et al. (1984a,b); K. Kawaguchi,
	ν_4; ν_5	DL	personal communciation; Tanaka et al. (1986); Ho et al.
	$\nu_1+\nu_4-\nu_4$, $\nu_1+\nu_5-\nu_5$		(1987); Amano & Tanaka (1986)
NH_3^+	ν_3	DF	Bawendi et al. personal communication[d]
SH_3^+	ν_1, ν_3; ν_2	DF; DL	Nakanaga & Amano (1987a, d); Amano et al. (1987)

more than four atoms

molecule	vibrational mode	comments	reference
HC_3NH^+	ν_1	DF	Lee & Amano (1987)

[a] Abbreviations: DF, difference frequency laser; DL, diode laser.
[b] B. D. Rehfuss, B. M. Dinelli, M. Okumura, M. G. Bawendi & T. Oka.
[c] M. Greubele, M. Polak & R. J. Saykally.
[d] M. G. Bawendi, B. D. Rehfuss, B. M. Dinelli, M. Okumura & T. Oka.

2. EXPERIMENTAL TECHNIQUES

The first high-resolution continuous-wave-difference-frequency spectrometer system was developed by Pine. Details of the system are described in his papers (see, for example, Pine 1974, 1976). Figure 1 shows a schematic diagram of a spectrometer system at National Research Council of Canada. Frequency-tunable infrared radiation is generated by mixing single-mode Ar^+ laser (Coherent INNOVA 90-5) radiation with tunable dye laser (Coherent 699-21) radiation in a $LiNbO_3$ crystal contained in a temperature-controlled oven. A typical power of the resulting infrared radiation was estimated to be about 30 μW at around 3 μm, with an input power of 400 mW of tunable dye laser radiation and 200 mW of single-mode Ar^+ laser radiation, by using the short circuit current responsivity of an InSb detector provided by the manufacturer. The linewidth of the difference-frequency radiation is a few megahertz. A combination of the two typical Ar^+ laser lines and a Rh-6G dye laser generates difference-frequency radiation in the range of *ca.* 2400–4400 cm^{-1}.

Figure 2 shows tuning characteristics of the difference frequency laser. This range covers most C–H, N–H, O–H, and S–H stretching vibrational bands. The wide frequency coverage

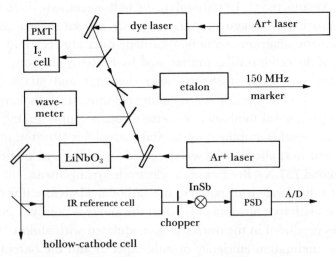

FIGURE 1. Schematic diagram of the difference-frequency laser system.

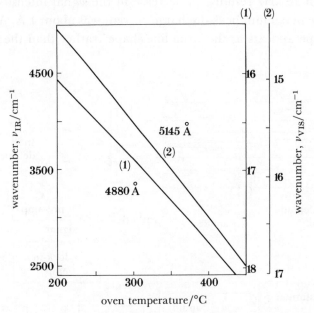

FIGURE 2. Tuning characteristics of LiNbO$_3$ crystal for (1) 488.0 nm and (2) 514.5 nm Ar$^+$ laser lines
(Amano 1983).

and easy tunability of difference frequency lasers are of great advantage in searching for lines
of unknown species. However, usually one encounters a severe overlapping problem from much
stronger lines, mostly because of parent molecules and stable reaction products, because most
of these stable species also have vibrational bands near the bands of interesting ions we want
to observe. So velocity modulation or amplitude-discharge modulation are essential to extract
the lines due to ions and moreover to enhance the detection sensitivity because the laser source
is very noisy. A hollow-cathode discharge, which is known to be an excellent source of ions, has
been successfully employed to observe ions in absorption with tunable laser sources first in
the far-infrared region (van den Heuvel & Dymanus 1982) and later in the 3–5 μm region

(Foster *et al*. 1984; Amano 1985). In particular, the hollow-cathode discharge combined with magnetic-field modulation (Kawaguchi *et al*. 1985) has proved to be a powerful technique.

Figure 3 is a schematic diagram of a hollow-cathode discharge cell and the signal detection system. The design of the cell is similar to that used by Foster *et al*. (1984). A cathode is made of copper or stainless steel pipe of 38 mm inner diameter and 80 cm in length. A set of multireflection mirrors is placed inside the vacuum enclosure. The discharge can be cooled by flowing liquid nitrogen, cooled methanol or water through copper tubing wound around the cathode. A sine-wave signal is amplified by an audio amplifier (Bryston model 4B, maximum output of 800 W), and is applied to the water-cooled stainless-steel anode through a step-up transformer (Hammond 737 X). Because of the electrode arrangements, the breakdown voltage is much higher when the anode is negative to the cathode. Therefore this type of cell works as a rectifier so that the discharge turns on only when the anode is positive, and the concentration of short-lived species produced in the discharge is modulated with almost 100 % efficiency. On the other hand, the modulation efficiency of stable species and the parent molecules is much less, depending on the modulation frequency and the lifetime of the species. It was found that the signal:noise ratio is optimum at a modulation frequency of around 5–20 kHz for ions. Modulation faster than 20 kHz resulted in decrease in the signal intensity. At a modulation frequency of 5 kHz, the maximum peak discharge current was about 1 A. After phase-sensitive detection, the line shape appears as the 'true line shape' rather than the first derivative.

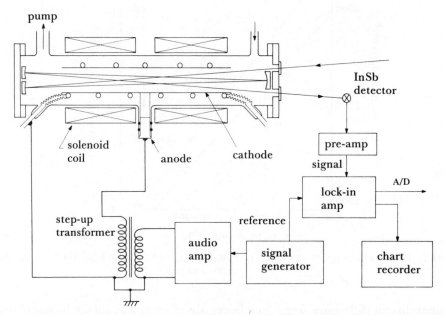

FIGURE 3. Schematic diagram of the hollow-cathode discharge cell. The cathode, which is made of either copper or stainless steel tubing, is 80 cm long and 38 mm in inner diameter. The typical optical pathlength is 24–32 m. The solenoid generates an axial magnetic field for examination of the field dependence of absorption lines.

The difference-frequency laser system is controlled by a microcomputer system (HP-9816). The scan of the ring dye laser is made in steps having increments of about 20 MHz. The infrared radiation passes through the multipass discharge cell and is detected by an InSb photovoltaic detector. The signal is sent to a lock-in amplifier, and the resulting DC signal is fed to an A–D converter. At the same time, the absorption signals of a suitable reference gas are recorded and sent to the computer memory through another A–D converter. A temperature-stabilized etalon (Burleigh CFT-500) in the dye laser beam with a free spectral range of 150 MHz is used to provide the frequency markers for interpolation of the frequency measurements. The marker signal is shaped and fed to a timer and pacer interface card. The signals are displayed on the screen with absorption lines of reference gas and with optional frequency markers. The frequency scale is calculated by using the markers and the frequencies of the reference lines, which are input by moving a cursor on the screen onto the line peaks and by typing their wave numbers. Then, a peak-finding routine is activated to obtain the wave numbers of the absorption lines of the sample of interest. The minimum detectable absorption of our system reaches about 0.002% or the minimum detectable peak absorption coefficient of 10^{-8} cm^{-1} with a 20 m pathlength with a time constant of 400 ms. More discussion on the sensitivity and the detectable abundance of ions is given by Nakanaga & Amano (1987c).

The amplitude-discharge modulation method does not always uniquely identify signals due to ions, but also detects lines due to short-lived neutral species. A hollow-cathode discharge, which is a good source of ions, is very attractive in spite of this shortcoming. Saito et al. (1985) demonstrated that an axial magnetic field of 100 G† applied to a hollow-cathode discharge resulted in almost 100% loss of the signal intensity in their detection of a millimetre-wave line of H_2D^+. It was found that, when the magnetic-field strength is increased, the intensity of the ion signals decreases rapidly, whereas the decrease of the signal intensity of neutral species is not so dramatic. This is contrary to the enhanced ion concentration in the extended negative glow in magnetic field found by DeLucia & co-workers (1983). An example of such magnetic field effect in a hollow-cathode discharge is given in figure 4. The different behaviour of ions under the axial magnetic field is used to discriminate signals due to ions from those of neutral species. This technique offsets the shortcomings to great extent, and it has been applied to the

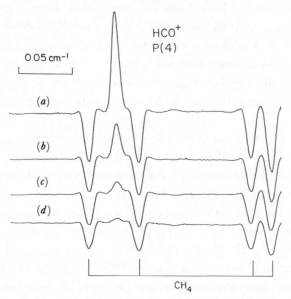

FIGURE 4. The P(4) line of the ν_1 fundamental band of HCO$^+$ together with the CH$_4$ lines recorded in a hollow-cathode discharge under several different axial magnetic fields: (a) 0 G; (b) 75 G; (c) 120 G; and (d) 150 G. The phase of the lock-in amplifier was set optimum for the HCO$^+$ line (Amano 1985).

† 1 G = 10^{-4} T.

detection of ions more or less as a standard technique with hollow-cathode discharges. In this context, it should be emphasized that the velocity modulation has an even greater advantage, because it can distinguish not only ions from neutrals but also negative ions from positively charged species (see, for example, Rosenbaum *et al.* 1986; Liu & Oka 1986).

As mentioned above, the frequency measurements rely greatly on known infrared lines of stable molecules measured with Fourier-transform spectrometers (a few typical examples are N_2O (Amiot & Guelachvili 1976; Amiot 1976), D_2O (Papineau *et al.* 1981) and H_2S (Lechuga-Fossat *et al.* 1984)) and with a difference-frequency laser system by Pine (1980*a,b*) (a few examples include H_2CO, C_2H_2 and H_2O). It is rather common that we are forced to use the spectra from different sources as references to measure a single band, and we have found that the internal consistency among the measured frequencies from different sources is very satisfactory.

3. RESULTS

(a) The ν_1 fundamental band of HOCS$^+$

HOCS$^+$ is isovalent with the ions HOCO$^+$ and HONN$^+$ studied recently in the submillimetre-wave and infrared regions. HOCO$^+$ was first (tentatively) identified in interstellar space through the observation of its millimetre-wave emission lines (Thaddeus *et al.* 1981). A definite confirmation was provided by Bogey *et al.* (1984) by detecting the submillimetre-wave lines in the laboratory. Shortly after, the ν_1 fundamental band of this ion was observed in a hollow-cathode discharge in the 3 µm region (Amano & Tanaka 1985*a,b*).

On the other hand, protonated N_2O was unknown spectroscopically when a search for the ν_1 band was undertaken. Vibration–rotation absorption lines were detected in a hollow-cathode discharge through a mixture of N_2O and H_2, and the lines were assigned as the ν_1 band of protonated N_2O (Amano 1986). However, it was not conclusive if the species detected was the O-protonated or N-protonated form from the experiment alone, although it was concluded that it was a bent molecule with an angle of HON or of HNN of about 117°. Two groups (Rice *et al.* 1986; Yamashita & Morokuma 1986) reported *ab initio* calculations immediately after and indicated that the O-protonated isomer is likely to be the species detected in the infrared experiment. Quite recently the submillimetre-wave spectrum was detected in the laboratory (Bogey *et al.* 1986*c*) guided by the molecular constants determined from the infrared analysis.

There has been no previous spectroscopic observation of protonated OCS. A few *ab initio* calculations have been performed to obtain the rotational constants, the proton affinity, and the relative stability of the three conceivable isomers (Taylor & Scarlett 1985; Scarlett & Taylor 1986). Protonated OCS has also been the subject of astrophysical interest (Fock & McAllister 1982; Taylor & Scarlett 1985). A possibility of the presence of HOCS$^+$ and HSCO$^+$ and the relative abundance to that of OCS in the interstellar clouds have been discussed by Fock & McAllister (1982). Taylor & Scarlett (1985) discussed possible assignments of unidentified lines observed in interstellar space to HOCS$^+$ or HSCO$^+$. However, no conclusions could be made because there was no spectroscopic study of these ions at the time.

Although an *ab initio* molecular-orbital (MO) calculation suggested that the S-protonated form, HSCO$^+$, may be more stable than the O-protonated, HOCS$^+$, we started the search for HOCS$^+$ in view of the fact that HSCN, an isomer of HNCS, is not known in the gas phase. Protonated OCS was generated in a hollow-cathode discharge in a gas mixture of H_2 (*ca.*

400 mTorr†) and OCS (*ca.* 15 mTorr) (Nakanaga & Amano 1987*c*). We found a series of absorption lines that were regularly spaced and could be assigned as the P- and R-branch lines of a-type transitions. Subsequently we found the b-type Q-branch lines, $^\mathrm{P}Q_1(J)$ and $^\mathrm{r}Q_0(J)$. Figure 5 shows these Q-branch lines.

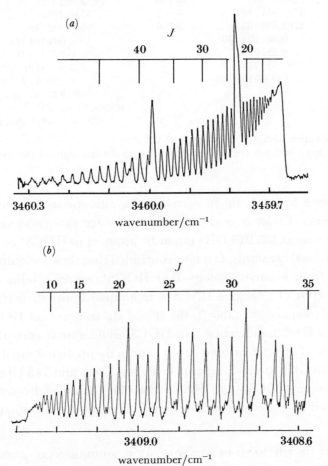

FIGURE 5. Examples of the Q-branch lines of $HOCS^+$ recorded with a detection time constant of 0.4 s. (Nakanaga & Amano 1987*c*). (*a*) The $^\mathrm{r}Q_0(J)$ transitions. The strong features around $J = 23$ and $J = 38$ are the lines of unidentified species. (*b*) The $^\mathrm{P}Q_1(J)$ transitions. The $J = 21$, $J = 31$, and $J = 32$ lines are overlapped with unassigned lines. An unassigned line is also evident between the $J = 33$ and $J = 34$ lines.

We have assigned about 550 lines (Nakanaga & Amano 1987*c*) and the molecular constants were determined by a least-squares fit with an effective hamiltonian in the *A*-reduced representation (Watson 1977). Table 2 lists the molecular constants determined from the fit. The mean deviation of the fit was 0.000 7 cm^{-1}. From the present infrared measurements, we can conclude that the species detected is an O-protonated isomer, because the vibrational frequency (3435 cm^{-1}) is very unlikely to be an S–H or C–H stretching vibrational frequency. Also the rotational constants obtained here are consistent with those calculated for $HOCS^+$ (Taylor & Scarlett 1985). The rotational constants are determined precisely to give predictions of the rotational transition frequencies with an accuracy of about several hundreds of kilohertz for the low-*J* lines and about a few megahertz for the lines in the 200–300 GHz region. These

† 1 mTorr ≈ 133.3 × 10^{-3} Pa.

TABLE 2. MOLECULAR CONSTANTS (IN MEGAHERTZ) OF $HOCS^+$

	ground state experiment	theory[a]	$\nu_1 = 1$ state experiment
A	782 695.7 (38)[b]	746 156	740 356.5 (40)
B	5750.551 (49)	5 803	5746.278 (47)
C	5703.030 (50)	5 758	5696.792 (49)
Δ_J	0.001 104 (20)		0.001 101 (19)
Δ_{JK}	0.281 9 (15)		0.263 1 (14)
Δ_K	993.68 (77)		233.30 (110)
Σ_J	0.000 019 8 (43)		0.000 012 7 (40)
Φ_K	0.0[c]		−46.823 (97)
ν_0			102 983 614.2 (27)
			[3435.163 611 (88) cm^{-1}]

[a] SCF calculation with (Taylor & Scarlett 1985).
[b] Values in parentheses are one standard error to the last digits of the constants.
[c] Fixed.

frequencies may prove to be useful in searching for this species in interstellar space or in laboratory experiments. Taylor & Scarlett (1985) suggested that two unidentified interstellar lines at 92.353 GHz and at 103.915 GHz might be assigned to $HOCS^+$ on the basis of their *ab initio* calculated rotational constants. It is now concluded that those assignments are not correct.

It was expected from a naive analogy that $HOCS^+$ would exhibit quasilinearity more conspicuously than $HOCO^+$, because HNCS is more quasilinear than HNCO (Winnewisser 1985). However, as compared in table 3, the A and Δ_K constants of $HOCS^+$ are very much smaller than those of HNCS, indicating that $HOCS^+$ is an almost normal bent molecule. An *ab initio* calculation by Taylor & Scarlett (1985) correctly predicted the A rotational constant for $HOCS^+$ to be 778 GHz at self-consistent field (SCF) level and 746 GHz with a correction. This difference between $HOCS^+$ and HNCS may be explained by considering resonance structures as discussed in a previous publication (Nakanaga & Amano 1987c).

TABLE 3. SUMMARY OF THE BAND ORIGIN FOR THE ν_1 FUNDAMENTAL BANDS AND THE MAJOR MOLECULAR CONSTANTS IN THE GROUND STATES OF $HOCS^+$ AND THE RELATED MOLECULES

	$HOCS^+$	HNCS[a,b]	HOCO$^+$[c,d]	HNCO[e,f]	HONN$^+$[g,h]	HNNN[i,j]
ν_0/cm^{-1}	3435.16	3538.6	3375.37	3537.7	3330.91	3336
A/MHz	782 696	1 361 530	789 939	918 504	625 954	610 996
B/MHz	5750.55	5883.46	10 773.72	11 071.01	11 301.56	12 034.15
C/MHz	5703.03	5845.61	10 609.44	10 910.58	11 084.28	11 781.45
Δ_K/MHz	993.7	57 520	1 123.6	6066	242.8	259.0

References: [a] Draper & Werner (1974). [b] Yamada *et al.* (1979). [c] Amano & Tanaka (1985b). [d] Bogey *et al.* (1986a, b). [e] Yamada (1972). [f] Yamada (1980). [g] Amano (1986). [h] Bogey *et al.* (1986c). [i] Dows & Pimentel (1955). [j] Bendtsen & Winnewisser (1975).

According to an *ab initio* calculation at SCF level (Scarlett & Taylor 1986), $HSCO^+$ is more stable than $HOCS^+$ by 34 kJ mol^{-1} (17 kJ mol^{-1} with inclusion of electron correlation effect). Although which is more stable seems to be a rather subtle problem, $HSCO^+$ may be detectable in the same discharge where $HOCS^+$ is found. So we searched for $HSCO^+$ in the range of 2450–2600 cm^{-1} by employing the same reaction conditions for $HOCS^+$. The search, however, revealed no evidence of $HSCO^+$.

(b) *Infrared spectra of* NH^+

NH^+ is a unique molecule in the sense that it has a low-lying electronic excited state $a^4\Sigma^-$. Optical emission spectra of NH^+ were first recorded by Lunt *et al.* (1935) more than 50 years ago. However, NH^+ has not been investigated so extensively as an isoelectronic diatomic molecule CH. Since the first high-resolution ultraviolet (uv) emission observation by Feast (1951), it has been known that the energy difference between the $X^2\Pi$ and $a^4\Sigma^-$ states is only 500 cm^{-1} (T_e), and the two states interact each other strongly. However, no quantitative analyses have been made to unravel the complicated perturbation.

The present high-resolution infrared investigation casts light directly on the perturbation. NH^+ was generated through a hollow-cathode discharge either in a mixture of N_2 (*ca.* 10 mTorr), H_2 (*ca.* 10 mTorr), and He (*ca.* 1.4 Torr) or in a mixture of NH_3 (*ca.* 10 mTorr) and He (*ca.* 1.4 Torr). Most measurements were done with dry-ice cooling, but ice or water cooling was equally effective with some change in the rotational temperature.

Because the perturbation is so severe, it is essential to treat the $X^2\Pi$ and $a^4\Sigma^-$ states simultaneously. The existing formulas for the rotation and fine-structure energy levels for $^4\Sigma$ electronic states (Budo 1937; Albritton *et al.* 1977) are extended to include the higher-order centrifugal distortion terms of the spin–spin and the spin–rotation interactions. The spin–orbit interaction between the two states that is a major cause of the perturbation is taken into account explicitly. The hamiltonian matrix was set up in terms of the case (*a*) basis. The details about the theory are given by Kawaguchi & Amano (1987a).

The search problem usually associated with unknown widely spaced vibration–rotation bands was greatly alleviated by using the term values given by Colin & Douglas (1968) to predict the transition frequencies. The assignments of the $^2\Pi$–$^2\Pi$ transitions were rather straightforward, because most lines were found near the predicted frequencies. A preliminary least-squares fit was made with the term values for extra levels observed by Colin & Douglas (1968) to predict the transitions within $a^4\Sigma^-$ and the ones between the $X^2\Pi$ and $a^4\Sigma^-$ states. The final least-squares fit was made with the measured infrared transitions, the combination differences, the term values obtained by Colin & Douglas (1968) corrected for systematic deviations (Kawaguchi & Amano 1987a), and the two far-infrared rotational transitions measured by Verhoeve *et al.* (1986) with appropriate relative weights.

Table 4 lists some major molecular constants for comparison with the previous experimental and theoretical values. Because Colin & Douglas (1968) did not fully analyse the perturbation quantitatively, the vibrational frequencies, the spin–orbit coupling constants, and T_0 value for the $a^4\Sigma^-$ state were not determined very well previously. It is not straightforward to derive B_e from B_0 and B_1, especially in a light hydride like NH^+. The equilibrium rotational constant for the ground state is estimated to be 470450 MHz after the corrections described by Kawaguchi & Amano (1987a), and the equilibrium bond length is calculated to be 1.069 Å†. As compared in table 4, the agreement between the experimental and theoretical values is very good except for an earlier scf calculation. The equilibrium bond length r_e in the $a^4\Sigma^-$ state was calculated without making the second-order correction, as it was expected to be small. Comparisons with several *ab initio* calculated values are also made in table 4.

† $1 \text{ Å} = 10^{-10} \text{ m} = 10^{-1} \text{ nm}$.

TABLE 4. MAJOR MOLECULAR CONSTANTS (RECIPROCAL CENTIMETRES UNLESS OTHERWISE STATED) OF NH^+

$X^2\Pi$ state

	experimental		calculated
B_0	15.331128 (46)[a]	15.35[b]	15.41 (PNO-CI)[c]
		15.587[d]	15.30 (CEPA)[c]
B_1	14.725118 (69)	—	14.80 (PNO-CI)[c]
			14.68 (CEPA)[c]
B_e	15.6925[e]	—	15.71 (PNO-CI)[c]
			15.61 (CEPA)[c]
D_0	0.0016313 (15)	0.00117[b]	
A_0	81.6568 (21)	77.8[b]	77.895[f]
ν_0	2903.1718 (21)	2922[b]	2938.6 (PNO-CI)[c]
		2980.65[d]	2893.0 (CEPA)[c]
r_e	1.069 Å		1.068 Å (PNO-CI)[c]
			1.071 Å (CEPA)[c]
			1.069 Å (SCF-CI)[g]
			1.070 Å (MRSDCI)[h]
			1.047 Å (SCF)[i]

$a^4\Sigma^-$ state

B_0	14.66001 (14)	14.69[b]	—
		14.984[d]	—
B_1	13.93295 (10)	—	—
B_e	15.0523		—
D_0	0.0018270 (33)	(0.0018)[b]	—
ν_0	2544.3004 (40)	2520[b]	—
		2652.29[d]	
T_0	323.8976 (45)	354[b]	
		189.9[d]	
r_e	1.092 Å	—	1.093 Å (SCF-CI)[g]
			1.079 Å (SCF)[i]

Notes: [a] One standard error to the last digits of the constants. [b] Colin & Douglas (1968). [c] Rosmus & Meyer (1977). [d] Experimental value of Colin & Douglas (1968) analysed by Wilson (1978). [e] Corrected, (see Kawaguchi & Amano (1987a)). [f] Wilson (1978). [g] Guest & Hirst (1977). [h] Kusunoki et al. (1986). [i] Liu & Verhaegen (1970).

(c) Laboratory measurements of the abundance ratio [HCO⁺]/[HOC⁺] and their astrophysical implications

The formation and depletion mechanism of HOC^+ and the abundance ratio [HCO⁺]/[HOC⁺] in interstellar space have attracted a great deal of interest theoretically and experimentally (see, for example, Dixon et al. 1984; Jarrold et al. 1986). Woods et al. (1983) attempted to detect this ion in interstellar space on the basis of a laboratory measurement by Gudeman & Woods (1982), and they tentatively assigned a line observed at the predicted frequency towards Sgr-B2 to HOC^+. Woods et al. (1983) estimated the abundance ratio [HCO⁺]/[HOC⁺] to be about 330, assuming the identification was correct, and discussed that it might be even more than 500 if the identification was incorrect.

There are two points that should be solved: (i) What is the branching ratio of the reactions

$$H_3^+ + CO \rightarrow HCO^+ + H_2, \tag{1}$$

$$H_3^+ + CO \rightarrow HOC^+ + H_2, \tag{2}$$

and (ii) What are the rate constants of depletion reactions of HOC^+? Illies et al. (1982, 1983) investigated the reaction system with the ion cyclotron resonance (ICR) mass-spectrometric technique. They concluded that $6 \pm 5 \%$ of the initial products from the reaction between H_3^+ and CO leads to formation of HOC^+, assuming the catalytic isomerization of HOC^+ by H_2,

$$HOC^+ + H_2 \rightarrow HCO^+ + H_2, \qquad (3)$$

is negligible. Later Wagner-Redeker et al. (1985) reinvestigated the system, and concluded that the 6% should be interpreted to be the lower limit, because the H_2 catalytic isomerization will occur to a substantial degree according to a phase-space calculation, which is contrary to another phase-space calculation by DeFrees et al. (1984) who concluded that the catalytic isomerization process (3) is too slow to explain the large depletion of HOC^+. Green (1984) also indicated that the isomerization was inefficient. More recently Jarrold et al. (1986) reinvestigated the reaction theoretically and experimentally. Their new conclusion was that the catalytic isomerization reaction (3) does not have a large activation energy contrary to the previous calculations (Green 1984; DeFrees et al. 1984), and that it is efficient process in depletion of HOC^+. Because H_2 is very abundant in dense interstellar clouds, the abundance ratio, $[HCO^+]/[HOC^+]$, tentatively given by Woods et al. (1983) may be too low, suggesting that the tentative identification of the line observed in Sgr-B2 might be in error.

Recently the infrared absorption of the ν_1 band of HOC^+ ($\nu_0 = 3268 \text{ cm}^{-1}$) was observed in a hollow-cathode discharge through a mixture of CO and H_2 (Nakanaga & Amano 1987b). The reaction conditions were quite different from those used in previous microwave observations (Gudeman & Woods 1982; Blake et al. 1983; Bogey et al. 1986b) in which a large amount of Ar is needed as a buffer gas. The relative intensity of the infrared absorption lines of HCO^+ and HOC^+ was measured as a function of the pressures of H_2 and CO. All the measurements were done with a fast-flow system and the pressure was measured at the exit of the cell. This method is direct and much more reliable than indirect and somewhat ambiguous mass-spectrometric measurements. From a rate-equation analysis, the rate constants of the key reactions were estimated with a reasonable accuracy.

First we found that the rotational temperature of HCO^+ and HOC^+ was approximately equal to the cell temperature ($-50\ ^\circ\text{C}$) from relative-intensity measurements of several vibration–rotation lines. All the measurements were done with a peak discharge current of 500 mA, unless stated otherwise. Details of the measurements and analysis are presented in a paper by Amano & Nakanaga (1987) and only a brief summary is given here.

The optimum CO pressure for the production of HOC^+ was found to be about 25 mTorr and it seems to shift towards higher pressure as the pressure of H_2 increases from 200 mTorr to 600 mTorr. The signal intensity of HOC^+ is less dependent on the pressure of H_2. In 1981, Nobes & Radom (1981) pointed out that the isomerization reaction

$$CO + HOC^+ \rightarrow HCO^+ + CO \qquad (4)$$

might be very efficient to deplete HOC^+. The pressure dependence measured here, indeed, suggests that CO plays a primary role in depletion of HOC^+ whereas H_2 has only a secondary effect.

Figure 6 shows an example of the absorption lines of HCO^+ and HOC^+ at a CO pressure of 15 mTorr and H_2 pressure of 420 mTorr. We obtained the intensity ratio $I(HCO^+)/I(HOC^+) = 88 \pm 3$ at this condition. The abundance ratio was calculated to be 382 ± 12 by using the transition dipole moments for the bands of HCO^+ ($\mu = 0.168$ D†) and HOC^+

† $1\text{ D} = 3.33 \times 10^{-30}\text{ C m}.$

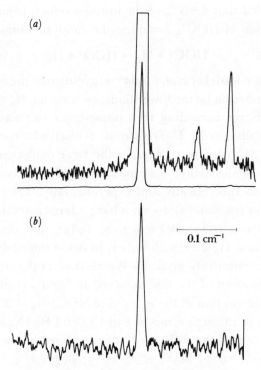

FIGURE 6. The R(8) lines of (a) HCO⁺ and (b) HOC⁺ recorded at the same gain settings with normalized infrared power. The middle trace was recorded with a hundredth of the gain of the top trace (Amano & Nakanaga 1987).

TABLE 5. REACTIONS CONSIDERED IN THE RATE-EQUATION ANALYSIS

reaction	rate constant
$H_2 + H_2^+ \rightarrow H_3^+ + H$	k_1
$H_3^+ + CO \rightarrow HCO^+ + H_2$	k_2
$H_3^+ + CO \rightarrow HOC^+ + H_2$	k_3
$HOC^+ + CO \rightarrow HCO^+ + CO$	k_4
$HOC^+ + H_2 \rightarrow HCO^+ + H_2$	k_5
$HA^+ + e \rightarrow products$	K_r

($\mu = 0.350$ D) recently calculated by P. Botschwina (personal communication). Because the rotational temperatures of HCO⁺ and HOC⁺ are found to be almost the same and the rotational constants of the two ions are similar, no further corrections have been made to derive the abundance ratio from the intensity ratio.

We have considered the reactions listed in table 5 to analyse the observed pressure dependence. A set of rate equations was set up, and the stationary concentrations of HOC⁺ and HCO⁺ were calculated to be

$$[HOC^+] = \frac{k_3[H_3^+][CO]}{k_4[CO] + k_5[H_2] + K_r[M]}, \tag{5}$$

$$[HCO^+] = \frac{[H_3^+][CO]}{K_r[M]} \left\{ k_2 + \frac{k_3(k_4[CO] + k_5[H_2])}{k_4[CO] + k_5[H_2] + K_r[M]} \right\}, \tag{6}$$

where

$$[H_3^+] = \frac{k_1[H_2][H_2^+]}{(k_2 + k_3)[CO] + K_r[M]} \tag{7}$$

[100]

and the abundance ratio $[HCO^+]/[HOC^+]$ is given as

$$\frac{[HCO^+]}{[HOC^+]} = \frac{k_2}{k_3} + \left(\frac{k_2}{k_3}+1\right)\frac{k_4[CO]+k_5[H_2]}{K_r[M]}. \tag{8}$$

In the derivations, the dissociative recombination rate K_r was assumed to be equal for H_3^+, HCO^+, and HOC^+. From (5) and (7), the optimum concentration of CO for formation of HOC^+ is given by

$$[CO]_m = \sqrt{\left\{\frac{K_r[M](K_r[M]+k_2[H_2])}{k_4(k_2+k_3)}\right\}}, \tag{9}$$

where [M] denotes the concentration of free electrons and all other possible impurities including the cell wall that lead to the depletion of the ions. It was found that the abundance ratio depended on both the CO and H_2 concentrations linearly. The abundance ratio $[HCO^+]/[HOC^+]$ extrapolated to the 'zero CO pressure' shows a linear dependence on the H_2 pressure. From the intercept and the slope, k_2/k_3 is found to be about 46 and $k_5/K_r[M]$ is approximately equal to 4.1×10^{-16} cm^3. The rate constant k_3 is obtained to be 3.9×10^{-11} cm^3 s^{-1}, if k_2 is 1.8×10^{-9} cm^3 s^{-1} (Jarrold et al. 1986). The slope of the plot of the abundance ratio $[HCO^+]/[HOC^+]$ against the CO pressure gives $k_4/K_r[M]$. However, we found that the plot is not completely linear, showing a systematic deviation from the model. So there is some ambiguity in determination of the slope. Thus we obtained

$$k_4/K_r[M] = (3.3\pm1.7) \times 10^{-15} \text{ cm}^3 \tag{10}$$

by taking the ambiguity into account. The optimum CO pressure for the formation of HOC^+ was found to be 25 mTorr when the H_2 pressure was 200 mTorr. Therefore from (9) we obtain

$$\sqrt{(k_2 k_4)}/K_4[M] = 2.2 \times 10^{-15} \text{ cm}^3 \tag{11}$$

by neglecting k_3. Equations (10) and (11) give $k_4 = (4.0\pm1.8) \times 10^{-9}$ cm^3 s^{-1} and $K_r[M] = (1.2\pm0.3) \times 10^6$ s^{-1}. Because the ratio $k_5/K_r[M]$ was already known above, k_5 is now determined as $(4.8\pm1.2) \times 10^{-1}$ cm^3 s^{-1}.

In the model stated above, the reverse reaction of (2) was neglected. However, it may not be negligible at least at the temperature range at which this experiment was done. The phase shift observed for the H_3^+ line as the CO pressure was increased may be an indication of the reverse reaction (Amano & Nakanaga 1987). The model used here may be over simplified. The most important species not incorporated in the model is probably CO^+. The reaction between CO^+ and H_2 is also thought to proceed rapidly (Jarrold et al. 1986).

The isomerization of HOC^+ by CO is a very fast process, and the rate constant k_4 is about an order of magnitude larger than that obtained by Wagner-Redeker et al. (1985). On the other hand, the catalytic isomerization process of HOC^+ to HCO^+ by H_2 is slower than that obtained by Jarrold et al. (1986). In the laboratory, CO plays a dominant role in isomerization of HOC^+. The rate constant of the formation reaction of HOC^+, k_3, was found to be approximately equal to that obtained by Jarrold et al. (1986).

In the laboratory, the abundance ratio $[HCO^+]/[HOC^+]$ was found to be 210 at the optimum reaction condition for production of HOC^+. This ratio, however, cannot be transferred to the astrophysical observations. In typical dense dark clouds, H_2 is much more abundant relative to CO. Therefore the rate of the catalytic isomerization is much faster than the isomerization by CO.

Let us assume some typical fractional abundances in dense molecular clouds to be $[CO]/[H_2] = 5.8 \times 10^{-5}$ and $[e]/[H_2] = 3.5 \times 10^{-8}$ (see, for example, Freeman & Millar 1983). Then $[HCO^+]/[HOC^+]$ is calculated to be of the order of 10^7 from (8). On the other hand, if a line of HOC^+ is detected, the abundance ratio may serve as a measure to determine the relative abundance of CO and free electrons. As stated above, all the measurements were done at about $-50\,°C$. Some preliminary measurements at a lower temperature, with liquid nitrogen for cooling the discharge, indicate no drastic change in abundance ratio $[HCO^+]/[HOC^+]$. However, the rate constants may be very different at the interstellar cloud temperature (see, for example, Adams & Smith 1987). If the isomerization of HOC^+ by CO proceeds via the complex as suggested by Nobes & Radom (1981), the process may be more efficient at lower temperature.

4. CONCLUDING REMARKS

The detection limit of ions has been pushed forward in recent years by about a factor of 10^3, which has been realized by the modulated hollow-cathode discharge or the velocity-modulation technique as well as by improvements in laser sources. Further improvement may be attained by employing the ratio technique, because the sensitivity is limited by the source noise in the current system. The infrared-spectroscopic techniques have proved to be powerful to allow the detection of such species as $C_2H_2^+$, $C_2H_3^+$, HC_3NH^+ and a number of negative ions that were regarded as a very remote goal only a few years ago.

In this article, a few examples were chosen from among the most recent results to illustrate several aspects of the high-resolution spectroscopy of ions. Although the accuracy of the frequency measurements is limited by the Doppler width, the molecular constants can be determined precisely to predict the pure rotational lines within a few megahertz uncertainty, because the number of lines obtainable from the analysis of the band spectrum is plentiful and fairly high-J transitions can be observed. This accuracy is good enough as a starting point to search for the microwave lines in the laboratories, and the observation of the microwave lines are essential to determine the rotational transition frequencies that lead to definite identification of the interstellar lines.

The high-resolution infrared absorption spectroscopy can be utilized for plasma diagnosis, as demonstrated by Pan & Oka (1987). In relatively simple reaction systems, the technique may be useful to identify the reaction intermediates, and more direct information on the reaction mechanism and the rate constants can be derived. Interactions with other fields such as gas-phase ion–molecule reaction kinetics, *ab initio* MO calculations, and molecular astrophysics will continue to be vital and productive.

I thank A. R. W. McKellar for reading the manuscript and for his comments.

REFERENCES

Adams, N. G. & Smith, D. 1987 In *Astrochemistry. IAU Symposium no. 120* (ed. M. S. Vardya & S. P. Tarafdar) Dordrecht: Reidel.

Albritton, D. L., Schmeltekopf, A. L., Harrop, W. L., Zare, R. L. & Czarny, J. 1977 *J. molec. Spectrosc.* **67**, 157.

Altman, R. S., Crofton, M. W. & Oka, T. 1984*a* *J. chem. Phys.* **80**, 3911.

Altman, R. S., Crofton, M. W. & Oka, T. 1984*b* *J. chem. Phys.* **81**, 4225.

Amano, T. 1983 *Bull. Soc. chim. Belg.* **92**, 565.

Amano, T. 1985 *J. opt. Soc. Am.* B **2**, 790.
Amano, T. 1986 *Chem. Phys. Lett.*, **127**, 101.
Amano, T. 1987 (In preparation.)
Amano, T., Kawaguchi, K. & Hirota, E. 1987 *J. molec. Spectrosc.* (In the press.)
Amano, T. & Nakanaga, T. 1987 *Astrophys. J.* (Submitted.)
Amano, T. & Tanaka, K. 1985*a* *J. chem. Phys.* **82**, 1045.
Amano, T. & Tanaka, K. 1985*b* *J. chem. Phys.* **83**, 3721.
Amano, T. & Tanaka, K. 1986 *J. molec. Spectrosc.* **116**, 112.
Amiot, C. 1976 *J. molec. Spectrosc.* **59**, 191.
Amiot, C. & Guelachvili, G. 1976 *J. molec. Spectrosc.* **59**, 171.
Bendtsen, J. & Winnewisser, M. 1975 *Chem. Phys. Lett.* **33**, 141.
Blake, G. A., Helminger, P., Herbst, E. & DeLucia, F. C. 1983 *Astrophys. J. Lett.* **164**, L69.
Bogey, M., Demuynck, C. & Destombes, J. L. 1984 *Astron. Astrophys.* **138**, L11.
Bogey, M., Demuynck, C. & Destombes, J. L. 1986*a* *J. chem. Phys.* **84**, 10.
Bogey, M., Demuynck, C. & Destombes, J. L. 1986*b* *J. molec. Spectrosc.* **115**, 229.
Bogey, M., Demuynck, C., Destombes, J. L. & McKellar, A. R. W. 1986*c* *Astron. Astrophys.* **167**, L13.
Brown, P. R., Davies, P. B. & Johnson, S. A. 1986 *Chem. Phys. Lett.* **132**, 582.
Budo, A. 1937 *Z. Phys.* **105**, 73.
Colin, R. & Douglas, A. E. 1968 *Can. J. Phys.* **46**, 61.
Crofton, M. W., Jagod, M.-F., Rehfus, B. D. & Oka, T. 1987 *J. chem. Phys.* **86**, 3755.
DeFrees, D. J., McLean, A. D. & Herbst, E. 1984 *Astrophys. J.* **279**, 322.
DeLucia, F. C., Herbst, E., Plummer, G. M. & Blake, G. A. 1983 *J. chem. Phys.* **78**, 2312.
Dixon, D. A., Komornicki, A. & Kraemer, W. P. 1984 *J. chem. Phys.* **81**, 3603.
Dows, D. A. & Pimentel, G. C. 1955 *J. chem. Phys.* **23**, 1258.
Draper, G. R. & Werner, R. L. 1974 *J. molec. Spectrosc.* **50**, 369.
Feast, M. W. 1951 *Astrophys. J.* **114**, 344.
Fock, W. & McAllister, T. 1982 *Astrophys. J. Lett.* **257**, L99.
Foster, S. C., McKellar, A. R. W. & Sears, T. J. 1984 *J. chem. Phys.* **81**, 578.
Freeman, A. & Millar, T. J. 1983 *Nature, Lond.* **301**, 402.
Green, S. 1984 *Astrophys. J.* **277**, 900.
Greubele, M., Polak, M. & Saykally, R. J. 1987*a* *J. chem. Phys.* **86**, 1698.
Greubele, M., Polak, M. & Saykally, R. J. 1987*b* *J. chem. Phys.* **86**, 6631.
Gudeman, C. S., Begemann, M. H., Pfaff, J. & Saykally, R. J. 1983 *Phys. Rev. Lett.* **50**, 727.
Gudeman, C. S. & Saykally, R. J. 1984 *A. Rev. phys. Chem.* **35**, 387.
Gudeman, C. S. & Woods, R. C. 1982 *Phys. Rev. Lett.* **48**, 1344.
Guest, M. F. & Hirst, D. M. 1977 *Molec. Phys.* **34**, 1611.
Ho, W.-C., Blom, C. E., Liu, D.-J. & Oka, T. 1987 *J. molec. Spectrosc.* **123**, 251.
Illies, A. J., Jarrold, M. F. & Bowers, M. T. 1982 *J. chem. Phys.* **77**, 5847.
Illies, A. J., Jarrold, M. F. & Bowers, M. T. 1983 *J. Am. chem. Soc.* **105**, 2562.
Jarrold, M. F., Bowers, M. T., DeFrees, D. J., McLean, A. D. & Herbst, E. 1986 *Astrophys. J.* **303**, 392.
Kawaguchi, K. & Amano, T. 1987*a* In *Symposium on Molecular Spectroscopy, Columbus, Ohio*, paper ME2.
Kawaguchi, K. & Amano, T. 1987*b* (In preparation.)
Kawaguchi, K. & Hirota, E. 1985 *A. Rev. phys. Chem.* **36**, 53.
Kawaguchi, K. & Hirota, E. 1986*a* *Chem. Phys. Lett.* **123**, 1.
Kawaguchi, K. & Hirota, E. 1986*b* *J. chem. Phys.* **85**, 6910.
Kawaguchi, K., Yamada, C., Saito, S. & Hirota, E. 1985 *J. chem. Phys.* **82**, 1750.
Kusunoki, I., Yamashita, K. & Morokuma, K. 1986 *Chem. phys. Lett.* **123**, 533.
Lechuga-Fossat, L., Flaude, J. M., Camy-Peyret, C. & Johns, J. W. C. 1984 *Can. J. Phys.* **62**, 1889.
Lee, S. K. & Amano, T. 1987 *Astrophys. J.* (Submitted.)
Lee, S. K., Amano, T., Oldani, M. & Kawaguchi, K. 1987 In *Symposium on Molecular Spectroscopy, Columbus, Ohio*, paper ME13.
Liu, D. J. & Oka, T. 1986 *J. chem. Phys.* **84**, 2426.
Liu, H. P. D. & Verhaegen, G. 1970 *J. chem. Phys.* **53**, 735.
Lunt, R. W., Pearse, R. W. B. & Smith, E. C. W. 1935 *Nature, Lond.* **136**, 32.
Nakanaga, T. & Amano, T. 1987*a* *Chem. Phys. Lett.* **134**, 195.
Nakanaga, T. & Amano, T. 1987*b* *J. molec. Spectrosc.* **121**, 502.
Nakanaga, T. & Amano, T. 1987*c* *Molec. Phys.* **61**, 313.
Nakanaga, T. & Amano, T. 1987*d* In *Symposium on Molecular Spectroscopy, Columbus, Ohio*, paper ME14.
Nobes, R. H. & Radom, L. 1981 *Chem. Phys.* **60**, 1.
Oka, T. 1980 *Phys. Rev. Lett.* **45**, 531.
Pan, F. S. & Oka, T. 1987 *Phys. Rev.* A **36**. (In the press.)
Papineau, N., Flaud, J. M., Camy-Peyret, C. & Guelachvili, G. 1981 *J. molec. Spectrosc.* **87**, 219.

Pine, A. S. 1974 *J. opt. Soc. Am.* **64**, 1683.
Pine, A. S. 1976 *J. opt. Soc. Am.* **66**, 97.
Pine, A. S. 1980*a* *M.I.T. Lincoln Laboratory*, report no. NSF/ASRA/DAR-78-24562.
Pine, A. A. 1980*b* *M.I.T. Lincoln Laboratory*, report no. F19628-80-C-0002.
Polak, M., Gruebele, M. and Saykally, R. J. 1987 *J. Am. chem. Soc.* **109**, 2884.
Rice, J. E. , Lee, R. J. & Schaefer, H. F. III 1986 *Chem. phys. Lett.* **130**, 333.
Rosenbaum, N. H., Owrutsky, J. C., Tack, L. M. & Saykally, R. J. 1986 *J. chem. Phys.* **84**, 5308.
Rosmus, P. & Meyer, W. 1977 *J. chem. Phys.* **66**, 13.
Saito, S., Kawaguchi, K. & Hirota, E. 1985 *J. chem. Phys.* **82**, 45.
Scarlett, M. & Taylor, R. 1986 *Chem. Phys.* **101**, 17.
Sears, T. J. 1987 *J. Chem. Soc. Faraday Trans.* **83**, 111.
Taylor, P. R. & Scarlett, M. 1985 *Astrophys. J. Lett.* **293**, L49.
Thaddeus, P., Guelin, M. & Linke, R. A. 1981 *Astrophys. J. Lett.* **246**, L41.
van den Heuvel, F. C. & Dymanus, A. 1982 *Chem. Phys.* **92**, 219.
Verhoeve, P., Ter Meulen, J. J., Meerts, W. L. & Dymanus, A. 1986 *Chem. Phys. Lett.* **132**, 213.
Wagner-Redeker, W., Kemper, P. R., Jarrold, M. F. & Bowers, M. T. 1985 *J. chem. Phys.* **83**, 1121.
Watson, J. K. G. 1977 In *Vibrational spectra and structure* (ed. J. R. Durig), vol. 5. Amsterdam: Elsevier.
Wilson, I. D. L. 1978 *Molec. Phys.* **36**, 597.
Winnewisser, B. P. 1985 In *Molecular spectroscopy: modern research* (ed. K. Narahari Rao), vol. 3. New York: Academic Press.
Woods, R. C., Gudeman, C. S., Dickman, R. L., Goldsmith, P. F., Huguenin, G. R., Irvine, W. M., Hjalmarson, A., Nyman, L.-A. & Olfsson, H. 1983 *Astrophys. J.* **270**, 583.
Yamada, K. 1972 Ph.D. dissertation, University of Tokyo.
Yamada, K. 1980 *J. molec. Spectrosc.* **79**, 323.
Yamada, K., Winnewisser, M., Winnewisser, G., Szalanski, L. B. & Gerry, M. C. L. 1979 *J. molec. Spectrosc.* **78**, 189.
Yamashita, K. & Morokuma, K. 1986 *Chem. Phys. Lett.* **131**, 237.

Phil. Trans. R. Soc. Lond. A **324**, 179–196 (1988)

Printed in Great Britain

High-resolution studies of autodetachment in negative ions

By K. R. Lykke, K. K. Murray, D. M. Neumark and W. C. Lineberger

Department of Chemistry and Biochemistry, University of Colorado, and Joint Institute for Laboratory Astrophysics, National Bureau of Standards and University of Colorado, Boulder, Colorado 80309-0440, U.S.A.

A review of high-resolution autodetachment spectroscopy of negative ions is given. The coaxial laser–ion-beam technique is used to probe the autodetachment dynamics of several molecular negative ions. The NH^- $(v = 1)$ ion is shown to decay mainly via vibrational autodetachment, whereas CH_2CN^- detaches via rotational autodetachment. The propensity rules implied by these systems are confirmed by PtN^-, which is found to decay by both routes.

Introduction

Before 1983, C_2^- and OH^- were the only high-resolution spectroscopically studied negative ions (Herzberg & Lagerqvist 1968; Lineberger & Patterson 1972; Hotop *et al.* 1974; Jones *et al.* 1980; Schulz *et al.* 1982; Hefter *et al.* 1983; Mead *et al.* 1985). Only within the last few years has there been a flurry of high-resolution spectroscopic studies of negative ions. There are two reasons for this delay. Although negative ions are seen throughout Nature (Massey 1976), they tend to occur in relatively low abundance so that it is very difficult to do absorption spectroscopy on any but the most easily made species. Also, the outer electron in negative ions is usually bound by less than *ca.* 1 eV so that there is very little 'room' in which to have another bound electronic state. Negative-ion spectroscopy has lagged behind neutral and positive-ion spectroscopy primarily because of these two inherent problems.

The conventional spectroscopic method is absorption of a beam of light by the absorbing species. Because negative ions occur in only low concentration, the sensitivity of absorption spectroscopy has had to be increased to observe negative ions in the gas phase. A major improvement in the sensitivity of absorption spectroscopy of negative ions (and positive ions) has come from the introduction of velocity-modulated spectroscopy (Saykally 1986). The negative ions that have been observed by this technique are typically among the most abundant negative ions in a plasma (Rosenbaum *et al.* 1986; Rehfuss *et al.* 1986; Kawaguchi & Hirota 1986; Tack *et al.* 1986) (OH^-, FHF^-, NH_2^-) and are down in concentration from positive ions by only *ca.* 1–3 orders of magnitude. A very desirable feature of direct absorption spectroscopy is that any molecule, in general, will have an absorption spectrum in some spectral region. Some drawbacks to this technique include limited ion-mass selectivity and large Doppler widths.

One method for circumventing the small concentration problem is to form a beam of negative ions and detect each photon absorbed with high efficiency. This can be done if each negative ion that absorbs a photon is excited to a state above the detachment continuum, and subsequently decays by autodetachment ('autodetachment spectroscopy'). Every electron and every neutral that is formed by this process can be detected with close to unity efficiency. Mass selectivity is another feature of this method, and by using a coaxial ion beam–laser beam, one is able to obtain greatly reduced Doppler widths. A drawback of autodetachment spectroscopy is that the upper state of the transition has to lie above the detachment threshold; no bound states can be probed by this technique.

A number of approaches have been utilized to circumvent the lack of negative-ion excited electronic states. One method that has been shown to be successful is vibrational autodetachment spectroscopy (Neumark *et al.* 1985). This involves excitation of the negative ion to a vibrationally excited state that lies above the detachment threshold. When applicable, this is a very good method for determining the structural characteristics of the ion and for obtaining detailed information on the interaction between the rotating and vibrating neutral and the detached electron. The crucial drawback to this technique is that it may not be general. The electron affinity of the neutral must lie below the vibrational level excited in the negative ion and the autodetachment lifetime must be in the range 10^{-7}–10^{-12} s for this approach to be successful.

Another method that might be general is two-photon detachment (one-photon resonant). This method has only been applied (Lineberger & Patterson 1972) in the case of C_2^-, in which both photons were the same frequency. The principle of the technique is to use a tunable laser to excite the molecular ion to an excited state (vibrational or electronic) and then detach this excited state with a high-power laser with photon frequency low enough so that it cannot detach ground-state molecules. Thus, molecules with electron affinities greater than their vibrational fundamental frequencies could be studied by this method. However, this method has not yet been demonstrated for vibrational spectroscopy.

A third method involves dipole bound states (DBS) of negative ions, states that are analogous to Rydberg states in neutral molecules. The spectroscopy of Rydberg states in neutrals has greatly increased the understanding of molecular structure (Gallas *et al.* 1985). There are an infinite number of Rydberg states in each molecule, bound by the $1/r$ potential between the positive-ion core and the electron, converging to the ionization limit of the neutral. Because there is no $1/r$ long-range potential between the neutral core and the electron in an anion, there are no Rydberg states in negative ions. However, a molecule with dipole moment greater than *ca.* 2 D† will bind an electron in the $1/r^2$ potential and produce Rydberg-type DBS (Crawford 1967; Garrett 1970, 1971, 1980, 1982; Turner 1977; Zimmerman & Brauman 1977*a, b*; Jackson *et al.* 1979, 1981; Mead *et al.* 1984; Lykke *et al.* 1984; Andersen *et al.* 1986, 1987; Marks *et al.* 1987). These states are weakly bound (typically less than or about 100 cm^{-1}) relative to the neutral molecule, the electron is bound in a very diffuse orbital, and the DBS is structurally similar to the neutral core, much like a Rydberg state. However, there are generally only one or two of these dipole-bound electronic states for each dipole neutral molecule, unlike the infinity of Rydberg states in each neutral molecule.

We can exploit these properties of DBS in negative-ion spectroscopy. Transitions from the ground electronic state of the negative ion to the DBS will yield autodetachment resonances in the upper, unbound rotational–vibrational levels. Analysis of this bound–quasibound transition will give structural information about both the anion ground state and the DBS, hence the neutral core. Also, because the electron is bound by the anisotropic $1/r^2$ dipole potential, the

† 1 D = 3.33×10^{-30} C m.

lifetime of the autodetaching resonances will give information about the interaction between a rotating–vibrating dipole and an electron. This will also help us to understand the process of autoionization in neutral molecules, where only a few systems have been studied with rotational resolution.

In this paper, we discuss three different autodetaching systems. The first one is NH^-, probed by vibrational spectroscopy. The next system discussed is CH_2CN^-, a dipole bound system studied by electronic spectroscopy. This species contains the major physical properties of DBS, but is the simplest to understand of the systems studied to date (Zimmerman & Brauman 1977 a, b; Jackson et al. 1979, 1981; Mead et al. 1984; Lykke et al. 1984; Marks et al. 1987). It looks very much like a linear dipole (CCN skeleton) with perturbing hydrogens. The last system discussed is PtN^-, which is studied by electronic spectroscopy and helps to elucidate many dynamical aspects of autodetachment. However, before these systems are discussed, a brief summary of the experimental methods will be given, followed by a brief review of the autodetachment process.

EXPERIMENTAL

An overview of the photodetachment apparatus is presented in figure 1. The basic idea is to merge a continuous tunable laser with a mass-selected, negative-ion beam. These two beams then interact for ca. 30 cm and the products of photoabsorption neutrals and electrons are observed. The ion beam is formed by extracting negative ions from any one of a number of sources (see below), accelerating the ions to 2–3 keV, mass selecting with a momentum analyser, and forming a well-collimated beam with lenses and apertures. Then the ion beam is turned 90° with a transverse electric field and copropagated with the laser beam before it is turned 90° again into a Faraday cup for monitoring. Typical ion currents are between 1 pA (ca. 10^7 ions s^{-1}) and 10 nA (ca. 10^{11} ions s^{-1}). The laser beam is sent through CaF_2 brewster windows and is monitored with Fabry–Perot spectrum analysers, a λ-meter for wavelength calibration and a power meter for cross-section normalization. The typical powers available are ca. 10–1000 mW in a ca. 0.1 cm^2 beam.

The remaining ions are deflected by the second transverse electric field and those neutral species formed by detachment (or dissociation) strike a CaF_2 plate. This impact produces secondary electrons that are then detected by an electron multiplier and counted to yield the total photodestruction cross section. The electrons that are detached in the interaction region are collected by a weak solenoidal field (ca. 5–10 G†) and detected by another electron multiplier.

The data are taken by scanning the laser in frequency (measured with a λ-meter) while monitoring the neutrals and electrons formed as a function of laser frequency and normalizing to the ion current and laser power. The various lasers that have been utilized in this photodetachment spectrometer are home-built single-mode cw dye lasers (covering the range from ca. 400–1000 nm), commercial F-centre lasers (F_A(II) and F_B(II) lasers covering the range from 2.2–3.5 μm) and home-built single-mode continuous-wave (cw) F_2^+-centre lasers (covering the range 1.5–1.8 μm).

† $1 G = 10^{-4} T$.

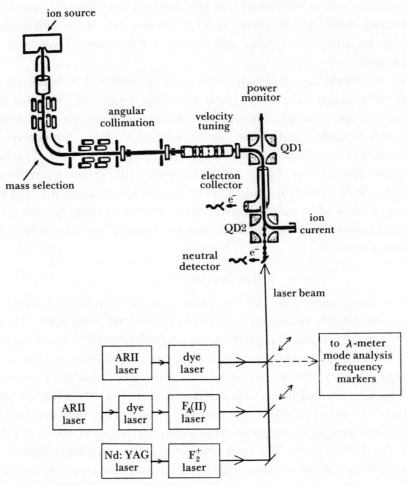

FIGURE 1. Overview of autodetachment apparatus. Note the wide selection of lasers required to cover a large portion of the spectrum from the visible to infrared.

AUTODETACHMENT PROCESSES AND RATES

Autodetachment mechanisms

Basically, the phenomenon of autodetachment (autoionization) stems from the breakdown of the Born–Oppenheimer approximation, namely, terms that couple the bound states to the continuum. From these couplings, we can calculate transition rates into the continuum via the Golden Rule (Berry 1966)

$$\text{rate} \propto \frac{2\pi}{\hbar} |\langle \psi_f | T | \psi_i \rangle|^2 \rho$$

$$\propto \frac{2\pi}{\hbar} k^{2l+1} |\langle \psi_f | T' | \psi_i \rangle|^2, \tag{1}$$

where T is the coupling operator, ψ_i is the wave function of the autodetaching negative ion state, ψ_f is the electron plus neutral wave function, and ρ is the density of states. The quantities k and l are the linear and angular momentum of the leaving electron and T' is the energy independent coupling operator. There are four different types of coupling operators of significance in molecular autodetachment (Berry 1966). These are: (1) configuration interaction (electron–electron); (2) spin–orbit; (3) vibrational–electronic; and (4) rotational–electronic. We outline each process briefly and discuss the most important types for autodetachment.

Configuration-interaction driven autoionization arises from electron–electron repulsion and has been studied in only a few systems (Fano 1961). This coupling will only induce autodetachment if the neutral electronic state lies below the negative-ion electronic state. In the systems discussed in this paper, the negative-ion electronic state lies below the neutral electronic state, so this form of coupling is unimportant.

Spin–orbit autoionization (autodetachment) has recently been studied in several systems (Andersen *et al.* 1986, 1987; Lefebvre-Brion *et al.* 1985, 1986) and occurs when a molecular core can make a spin-orbit change to a lower state to allow the electron to leave the molecule. Spin-orbit autodetachment is unimportant for the molecules discussed here.

Rotational and vibrational autodetachment processes stem from Born–Oppenheimer breakdown (i.e. a coupling between nuclear motion and electronic motion). The coupling terms are (Hefter *et al.* 1983)

$$T' = T_{\mathrm{vib}} + T_{\mathrm{rot}}$$

$$= -\frac{\hbar^2}{2\mu}\frac{\partial}{\partial R}\bigg|_{\mathrm{el}}\frac{\partial}{\partial R}\bigg|_{\mathrm{vib}} - \frac{\hbar^2}{2\mu R^2}N|_{\mathrm{el}}\,N|_{\mathrm{rot}}. \tag{2}$$

The second term is a coupling similar (Berry 1966) to Λ-type doubling and produces coupling matrix elements proportional to $[N(N+1)]^{\frac{1}{2}}$, a coupling that should yield autodetachment rates proportional to the rotational energy. The first term in (2) can be shown (Neumark *et al.* 1985) to be proportional to $1/\epsilon_{\mathrm{if}}$ where ϵ_{if} is the nuclear energy (i.e. energy of rotation and vibration) given up in the transition. The autodetachment rate for the process primarily driven by this term would thus be proportional to $1/(\epsilon_{\mathrm{if}})^2$. Both vibrational and rotational autodetachment are important for the systems discussed in this paper.

Autodetachment line shapes

Fano (1961) has given the energy dependence of the autoionization cross section for a discrete state interacting with a continuum. The cross section

$$\sigma(\epsilon) = \sigma_{\mathrm{b}} + \frac{\sigma_{\mathrm{a}}(\epsilon+q)^2}{(\epsilon^2+1)} \tag{3}$$

depends on $\epsilon = (E-E_0)/\frac{1}{2}(\hbar\Gamma)$, the reduced energy variable, and q, the asymmetry parameter. The quantity q^2 is the ratio of autodetachment to direct detachment. In the limit of $q \to \infty$, the resonance line approaches a lorentzian with a full width at half maximum (FWHM) given by Γ and is simply related to the lifetime of the state by the Heisenberg Uncertainty Principle.

Most molecular negative ions possess many (*ca.* 1000–10000) different initial vibration–rotation states each of which undergoes direct photodetachment with a cross section *ca.* 10^{-19} cm^2. Therefore, for the autodetachment to be observed on this background, the cross section for autodetachment must be greater than about 10^{-16} cm^2. Also, this autodetaching state cannot interact with all of the underlying continua so that $q > 100$, which fulfills the above criterion for a lorentzian. Therefore, by fitting the resonances (at least in the vibrational ground state in the autodetaching states) to lorentzians, the lifetime (and the rate) is determined.

EXAMPLES OF NEGATIVE-ION AUTODETACHMENT

(a) NH⁻

The only detailed example of vibrational autodetachment from a ground electronic state available at present is NH^-. The lack of other examples can in part be traced to the difficulty of finding systems whose vibrational fundamental frequencies exceed the electron affinity, and the lack of convenient, high-power sources of tunable infrared radiation. Shown in figure 2a is a simplified overview of the relevant NH^- and NH energy levels. The first-excited vibrational level of NH^- is unbound with respect to detachment to the lowest vibrational level of NH. Therefore, the vibration–rotation spectrum of NH^- should appear as an autodetachment spectrum riding on a small non-resonant detachment continuum. Neumark *et al.* (1985) observed transitions up to $J' \approx 13.5$ in $v' = 1$, and Al-za'al *et al.* (1986) extended these data to $J' \approx 36.5$ in $v' = 1$.

The major findings in Neumark *et al.* (1985) were a set of spectroscopic constants (which were extremely close to the calculated (Rosmus & Meyer 1978) values), and the dependence of the autodetachment rate on rotational quantum number. We will not go into any of the spectroscopic findings here but will concentrate on the autodetachment dynamics in NH^-.

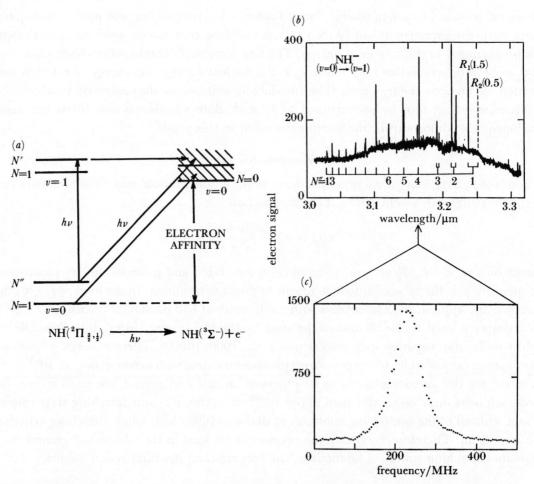

FIGURE 2. (a) Simplified energy-level diagram of the lowest vibrational levels in NH^-. (b) Low-resolution autodetachment spectrum of NH^- between 3.0 and 3.3 μm. (c) High-resolution (*ca.* 20 MHz) spectrum of one of the lines in NH^-.

The NH⁻ ion is isoelectronic with the well-studied molecule OH. Therefore, the ground electronic state is $X^2\Pi_{\frac{1}{2},\frac{3}{2}}$ and the vibrational spectrum will consist of P-, Q- and R- branches, each composed of several sub-branches (Herzberg 1950). The important point for our discussion is that there are four different fine-structure levels for each rotational level in both $v = 0$ and $v = 1$. These levels are caused by spin–orbit splitting in the $^2\Pi$ state $[^2\Pi_{\frac{1}{2}}(F_2)$ and $^2\Pi_{\frac{3}{2}}(F_1)]$ and further split into Λ-doublet levels in each rotational level. The spin decouples from the molecular axis toward high J and recouples to the rotational axis to produce F_1 components $(J = N+1)$ and F_2 components $(J = N-1)$ instead of the case (a) components in the $^2\Pi_{\frac{3}{2}}$ and $^2\Pi_{\frac{1}{2}}$ manifolds, respectively.

A broadband scan of NH⁻ with the F-centre laser is shown in figure $2b$ with a single-mode scan shown in figure $2c$ (this is an expansion of $ca.$ 20000 over figure $2b$). This expanded view shows that the resonant part of the coupling dominates the direct photodetachment process, so that the observed line shape is lorentzian, and that the observed line widths are simply related to lifetimes. Figure $3a$ shows the observed linewidths of the autodetaching transitions as a function of rotational quantum number N for the upper Λ-doublet level (upper graph) and the lower Λ-doublet level (lower graph). Note the change in scale on the graphs. The resolution of this experiment is $ca.$ 30 MHz and does not separate the hyperfine components, so we can only state an upper limit on linewidth of 30 MHz and thus a lower limit on the lifetime of $ca.$ 5 ns.

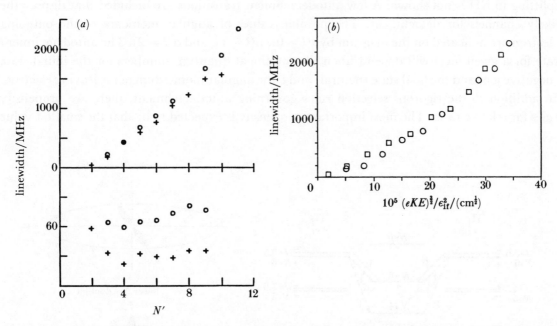

FIGURE 3. (a) Autodetachment rate (linewidth) against rotational quantum number for NH⁻ $(v = 1)$. The linewidths were determined by fitting the resonances to lorentzians. The upper (lower) graph corresponds to the upper (lower) Λ-doublet levels. (b) Autodetachment rates for the upper Λ-doublet levels in NH⁻ $(v = 1)$ plotted in a reduced energy release unit. If vibrational autodetachment were the only driving force, the data points would fall on a straight line in this representation. Symbols: $\square$, F_1; $\bigcirc$, F_2.

The important points to notice in figure $3a$ are the following: (a) a relative unimportance of the autodetachment rate on spin-orbit component; (b) a drastic dependence on Λ-doublet level; and (c) an approximately linear increase of autodetachment rate as a function of the rotational quantum number for the upper Λ-doublet levels. In the following we explain these phenomena with simple pictures, but first we need to see which rotational channels are open for detachment.

NH^- $(v = 1)$ decays mainly via vibrational and rotational autodetachment. This should be a rapid process because of the $\Delta v = -1$ propensity rule developed by Berry (1966) for an ion and neutral with similar potential-energy curves. However, the relative energy between NH^- $(v = 1)$ and $NH(v = 0)$ is not known exactly, so there is a corresponding uncertainty in the energy of the outgoing electron and even some uncertainty as to which rotational levels of NH are open channels for detachment. Neumark *et al.* (1985) observed transitions to the lowest rotational level of $NH^-(v = 1)$ (Λ-doubling was not resolved) so the electron affinity must be less than this value (*ca.* 0.374 eV), but still within the error bars quoted by Engelking & Lineberger (1976) (0.381 ± 0.014 eV). This gives the electron affinity as 0.370 ± 0.004 eV and implies that the autodetached electron from the lowest rotational level in $NH^-(v = 1)$ must have less than 0.008 eV (*ca.* 65 cm^{-1}) of energy.

NH^- approaches Hund's case (*b*) very rapidly ($J \approx 3.5$ is already in the transition to case (*b*)) so that the spin–orbit component does not have a substantial effect on the autodetachment rate; at high Js, the rotational levels approach the same energy for a given N. This transition is evident in figure 4*a*, which depicts a few representative levels for NH ($v = 0$, in the centre) and NH^- ($v = 1$, on the outside), drawn from an electron affinity of *ca.* 0.370 eV. The Λ-doublet splitting in NH^- is shown greatly exaggerated for clarity, but the very small spin splitting in NH is not shown. A few autodetachment transitions are included that display the likely channels for detachment. The possible values of angular momenta of the outgoing electron are indicated on the diagram by $s(l = 0)$, $p(l = 1)$, and $d(l = 2)$. The autodetachment rate for a given level will depend upon the rotational quantum numbers for the initial state (negative ion) and the final state (neutral) and l, the angular momentum of the leaving electron. In addition to the rigorous selection rules governing autodetachment, there are propensity rules for relative rates. The most important propensity is expected to be that the smallest value

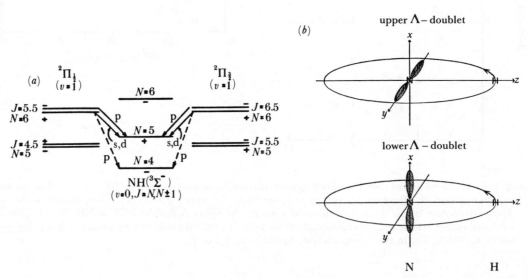

FIGURE 4. (*a*) Relevant energy levels for NH^- and NH showing some possible autodetachment transitions. (*b*) Orbital lobes involved in detachment from NH^-. Detachment occurs from the doubly occupied p-orbital localized on the N atom, almost directly on the centre of mass. The plane of rotation is the y–z plane, and the J-vector in the high rotational states lies along the x-axis.

of l possible for the parities of the initial and final states will be the strongest (fastest) transition. This is a result of the centrifugal barrier *ca.* $[l(l+1)]/r^2$, which the outgoing electron must conquer to leave the neutral. The next most important propensity is that rotational transitions with the smallest ΔN possible are favoured because the outgoing electron cannot exert much torque on the neutral core.

Neumark *et al.* (1985) concluded that the transitions marked with solid lines in figure 4*a* were the most probable, based upon a rough measurement of the kinetic energy of the detached electron. They could not obtain exact energies, but they could determine that both Λ-doublets of each spin–orbit manifold detached to the same level of NH. However, they observed detachment only up to $J \approx 13.5$ because of the limited user time on the borrowed laser. If one extrapolates the energy levels in figure 4*a* to higher Js, one finds that at some point the $\Delta N = -1$ channel becomes endothermic (at *ca.* $N = 29$ in NH^-) and leaves $\Delta N \geqslant -2$ as the smallest change energetically possible. This is because the difference in the rotational B, D values (e.g. $N = 35$ for NH lies *ca.* 200 cm^{-1} above $N = 36$ for NH^-). Al-za'al *et al.* (1986) reported observing up to $J \approx 36.5$ in $v = 1$ of NH^-. They report only the lower Λ-doublet because the upper Λ-level has broadened so severely that it cannot be distinguished from the underlying photodetachment continuum. If the $\Delta N = -1$ transition becomes unavailable, one should observe a narrowing of the linewidth for the upper Λ-level unless there is another factor influencing the relative detachment rates, other than low ΔN and low l. We will now attempt a brief explanation for this other factor.

NH^- and NH have the following valence molecular orbitals:

$$NH^-(^2\Pi):(2p\sigma)^2(2p\pi_{-1})(2p\pi_{+1})^2$$
$$NH(^3\Sigma^-):(2p\sigma)^2(2p\pi_{-1})(2p\pi_{+1}).$$

Therefore detachment occurs from the doubly occupied $p\pi$ orbital lobes that will tend to be oriented in different directions relative to the plane of rotation for the different Λ-levels (Andresen & Rothe 1985; Helm *et al.* 1982). The lower part of figure 4*b* corresponds to the detached p-electron lying out of the plane of rotation (lower Λ-doublet level) and the upper part of figure 4*b* corresponds to the detached p-electron (i.e. the doubly occupied orbital) lying in the plane of rotation (upper Λ-doublet level). Because the centre of mass for NH^- lies almost on the nitrogen nucleus, the p-orbital out of the plane of rotation will not feel much torque toward higher J, whereas the p-orbital in the plane of rotation will feel a large force. So rotation will tend to decouple the electron much more readily in the upper Λ-doublet state, in agreement with experiment.

Now we can make the autodetachment rate for NH^- a little more quantitative. As explained in the previous section, the autodetachment rate is given by Fermi's Golden Rule

$$W = \frac{2\pi}{\hbar}|\langle\psi_f|T|\psi_i\rangle|^2\rho. \tag{4}$$

We calculate the vibrational autodetachment rate of the upper Λ-doublet of NH^- by following the same line of reasoning as in Neumark *et al.* (1985). However, there is a factor of k^{2l} (i.e. $(eKE)^l$) that was ignored by Neumark *et al.* This factor arises from the tunnelling motion of the leaving electron through the $l = 1$ angular-momentum barrier. Therefore, a plot of Γ (autodetachment rate) against $(eKE)^{\frac{3}{2}}/(\epsilon_{if})^2$ (figure 3*b*) should yield a straight line if vibrational autodetachment is the driving force behind detachment from the upper Λ-doublet levels.

17-2

Figure 3b shows that vibrational autodetachment is indeed the driving force behind detachment. However, there is noticeable deviation from a straight line for large releases, implying that rotational autodetachment is at work in the upper levels.

(b) CH_2CN^-

The CH_2CN^- ion was chosen for study because its electron affinity (ca. 1.5 eV) (Marks et al. 1986; Moran et al. 1987) makes the photodetachment threshold region readily accessible to cw single-mode lasers, and the CH_2CN dipole moment (ca. 3D) should support a weakly bound CH_2CN^- dipole-bound state. Indeed, early low-resolution photodetachment studies by Zimmerman & Brauman (1977 a, b) showed unresolved structure near the photodetachment threshold, structure that was interpreted as confirming the existence of the DBS.

The total photodetachment cross section for CH_2CN^- with the dye laser in its broadband ($\Delta \nu \approx 1\ cm^{-1}$) configuration is shown in figure 5 a. This is similar to the spectrum obtained by Marks et al. (1986). The spectrum shows sharp structure due to autodetachment,

$$CH_2CN^- + h\nu \rightarrow (CH_2CN^-)^* \rightarrow CH_2CN + e^-, \tag{5}$$

where $(CH_2CN^-)^*$ is the dipole-bound state. This structure is superimposed on a smoothly rising background due to direct photodetachment,

$$CH_2CN^- + h\nu \rightarrow CH_2CN + e^-, \tag{6}$$

consisting of unresolved transitions to states in the (neutral $+ e^-$) continuum. The spectrum shows two vibronic bands. The autodetachment structure in the vicinity of $12400\ cm^{-1}$ is caused by transitions between the vibrationless levels of the two electronic states of CH_2CN^-, and the structure near $12000\ cm^{-1}$ consists of hot-band transitions in which the ground electronic-state ions are in the $v = 2$ level of the hydrogen out-of-plane bending mode. The structure in this low-resolution spectrum represents unresolved Q-branches and is characteristic of a perpendicular electronic transition for a slightly asymmetric prolate rotor (Herzberg 1967).

A low-resolution spectrum was obtained for CD_2CN^- as well, and spectra covering the 0–0 and hot-band regions of CH_2CN^- and CD_2CN^- were taken with the laser in its single-mode configuration. Approximately 10 000 transitions were recorded. A single-mode scan of a Q-branch is shown in figure 5 b. The individual rotational transitions are clearly resolved, and the high J-transitions (at higher frequency) are considerably broader than the low J-transitions. This indicates that autodetachment is faster for the higher rotational levels of the dipole-bound state.

The observed transitions were fit to an asymmetric-top hamiltonian, including centrifugal distortion, and the complete set of constants thereby obtained for both isotopes will be listed in a future publication (Lykke et al. 1987). Table 1 gives A, B, and C for the $v = 0$ levels of the ground and dipole-bound states of CH_2CN^-. From these constants we see the ion is nearly a symmetric prolate rotor with $A \gg B \approx C$, so each rotational level can be labeled by J, the total rotational angular momentum, and the nearly good quantum number K, the projection of J along the a-axis of the ion (the C—C≡N axis). The inertial defect Δ obtained from the rotational constants indicates that the dipole-bound state is planar and the ground state is either planar or has a very small barrier to inversion. A small inversion barrier has been inferred from a recent photoelectron-spectroscopy study of CH_2CN^- (Moran et al. 1987).

[114]

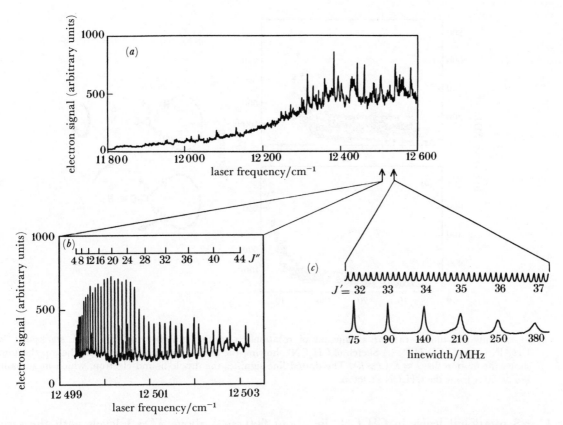

FIGURE 5. (a) Low-resolution (ca. 1 cm^{-1}) photodetachment spectrum of CH$_2$CN$^-$, showing the resonant auto-detachment structure superimposed on the direct photodetachment background. (b) A high-resolution (ca. 30 MHz) expansion (× 200) of the indicated portion of (a) (^{r}Q$_3$). (c) High-resolution scan showing the variation in linewidth as a function of J for a certain K, with ca. 250 MHz Fabry–Perot pips as calibration.

TABLE 1. PARTIAL LIST OF ROTATIONAL CONSTANTS FOR $v = 0$ LEVELS OF GROUND AND DIPOLE-BOUND ELECTRONIC STATES OF CH$_2$CN$^-$

	ground state	dipole-bound state
A/cm^{-1}	9.29431 (14)	9.51035 (17)
B	0.338427 (20)	0.341049 (21)
C	0.327061 (21)	0.328764 (21)
Δ/(uÅ^2)	−0.0827	0.0744

Finally, an analysis of the asymmetry doubling for the low K sub-bands shows that this is a $^1B_1 \leftarrow {}^1A_1$ electronic transition, and that the symmetry of the dipole-bound orbital in the upper state is a_1. Bearing in mind that this orbital is localized on the positive end of the neutral CH$_2$CN core, figure 6c qualitatively depicts the shape of the orbital for the dipole bound electron.

The linewidths of the transitions are plotted as a function of J' for different values of K' in figure 6a for CH$_2$CN$^-$ and figure 6b for CD$_2$CN$^-$. Here J' and K' are the rotational quantum numbers for the dipole-bound state. The striking feature of these plots is the abrupt rise in the autodetachment rates near $J' = 33$ for CH$_2$CN$^-$ and $J' = 38$ for CD$_2$CN$^-$, an onset that is independent of K'. This onset is not caused by total rotational energy because, for example,

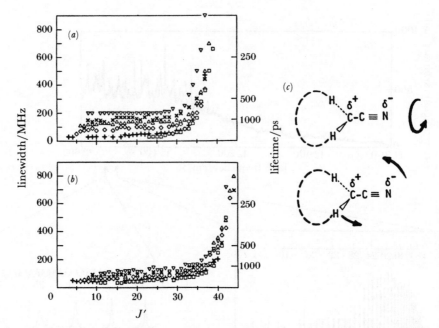

FIGURE 6. Autodetachment rate as a function of rotational quantum number for various K stacks in: (a) CH_2CN^- and (b) CD_2CN^-. (c) Sketch of CH_2CN^- showing the limiting rotational motions. The top (bottom) shows the motion for $J \approx K$ ($J \gg K$). The dotted line signifies the dipole-bound electron, which in actuality lies *ca.* 30 Å from the CH_2CN skeleton.

$K' = 8$ rotational levels in CH_2CN^- lie about 600 cm^{-1} above $K' = 1$ levels with the same J'.

The various mechanisms responsible for autodetachment in negative ions and autoionization in neutrals have been discussed previously. In this case, autodetachment is occurring from the vibrationless rotational levels of the dipole bound state, and rotational autodetachment should be the dominant mechanism. The rotating ion must undergo a rotational de-excitation, and the energy thereby released must eject the dipole-bound electron. In general, autodetaching transitions involving small values of ΔJ and l are favoured.

The autodetaching transitions energetically available to a given rotational level of the dipole-bound state can be determined readily once one knows the binding energy of the outermost electron in the dipole-bound state; this is the energy difference between the rotationless levels of the dipole-bound state and the ground state of neutral CH_2CN. Because the rotational levels of the dipole-bound state lying below the detachment threshold cannot autodetach, an upper limit to the binding energy can be obtained from the missing lines at low J in the spectra. For example, no transitions to $K' = 2$, $J' < 8$ are observed, placing an upper limit of *ca.* 60 cm^{-1} on the binding energy. With this limit, it can be shown that the rapid increase in the autodetachment rates in figure 6a, b coincides with an autodetaching transition of $\Delta J = -2$ or -3, $\Delta K = 0$, becoming energetically accessible. (The exact value of ΔJ depends on the binding energy, which may be lower than 60 cm^{-1}.)

For lower values of J' where the autodetachment rate is slow, autodetachment can only occur by a larger ΔJ transition, which is expected to be slow, or by a transition in which K as well as J changes. From figure 6c, one can qualitatively see why an autodetaching transition requiring a change in K might be very slow. Rotational autodetachment requires coupling between rotational and electronic motion, and rotational angular momentum about the a-axis is going to be poorly coupled to the diffuse orbital of the dipole-bound electron, which is nearly cylindrically symmetric about the a-axis.

The strong dependence of autodetachment rate upon rotation has been observed in other cases involving dipole-bound states. In each system, abrupt rate increases can be correlated with the availability of small ΔJ, l autodetaching transitions. This type of autodetachment rate behaviour appears to be a signature of a dipole-bound state. The next system reported here, PtN$^-$, makes the importance of small ΔJ changes unambiguous.

(c) PtN$^-$

Molecules containing open d-shell atoms are expected to support many negative-ion electronic states. PtN$^-$ is such a system, one that was accidentally discovered in 1972 to contain resonances. Hotop & Lineberger (1973) observed sharp resonances in the photodetachment cross section of PtN$^-$ when they were studying threshold photodetachment of Pt$^-$, Au$^-$ and Ag$^-$. The cold-cathode sputter source that produced the Pt$^-$, Au$^-$ and Ag$^-$ also produced copious quantities of PtN$^-$ when N_2 was used as the sputtering gas. Because of their low-resolution detachment apparatus ($ca.$ 3 cm^{-1}), they were not able to resolve rotational structure, but did set a lower limit on the lifetime of $\tau \gtrsim 5 \times 10^{-13}$ s. Because their apparatus only measured the photodestruction of the negative ion, they were unable to determine if the excited states of PtN$^-$ decayed by autodetachment or predissociation.

Murray et al. (1987) recently studied the same PtN$^-$ resonances with much higher resolution ($ca.$ 30 MHz) and were able to resolve rotational structure, isotopic structure and hyperfine structure. Figure 7 shows the structure observed in their experiment and proves that the resonances observed by Hotop & Lineberger (1973) are caused by autodetachment. Figure 7a is the broadband photodetachment spectrum of PtN$^-$ and is very similar to figure 9 in Hotop & Lineberger. A 'blow-up' of a small region taken with the same resolution ($ca.$ 0.5 cm^{-1}) is shown in figure 7b and contains some rotationally resolved structure. Figure 7c is a high-resolution ($ca.$ 30 MHz) scan of one of the resonances and shows isotopic splitting due to the three most abundant Pt isotopes. Also, the hyperfine splitting due to the ^{195}Pt nucleus appears at this resolution.

The other peaks in figure 7a were scanned in medium-to-high resolution but no other narrow resonances were observed. Some resonances $ca.$ 10 GHz wide were found around 15400 cm^{-1}, but these were too broad and congested to analyse. A brief discussion of the analysable part of the spectrum follows.

The data obtained for the band system near 16400 cm^{-1} yielded to a spectroscopic analysis by using combination differences, followed by a least-squares fit. The resonances are from a $^1\Pi$–$^1\Sigma$ transition, followed by autodetachment from the $^1\Pi$ levels. The initial combination difference fit was hampered by the fact that Js lower than $ca.$ 45 were too broad to distinguish from direct detachment and Js above $ca.$ 59 were too narrow to observe (these were too long lived and did not detach in the apparatus).

The autodetachment rates against rotational quantum number are shown in figure 8. As is immediately evident, PtN$^-$ has a much different detachment-rate dependence from the other negative ions studied to date. The autodetachment lines are very broad (rapid detachment rate) at low J and very narrow (very long lived) at high J. This is in spite of the fact that $J = 57$ lies $ca.$ 450 cm^{-1} above $J = 46$. Notice also that the Q-branch has a slightly different

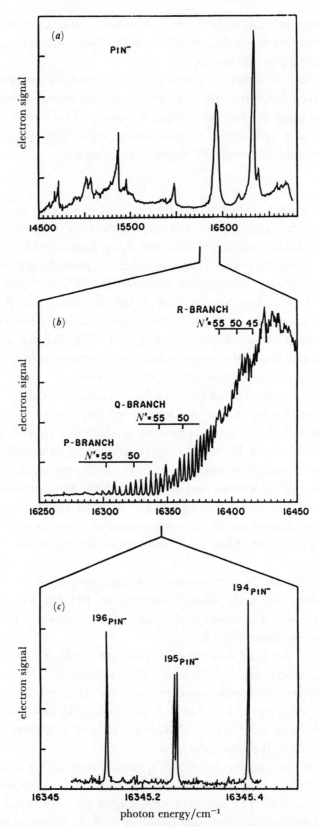

FIGURE 7. (a) Low-resolution photodetachment spectrum of PtN⁻. (b) Expansion of the indicated region of (a), showing a hint of resolved structure. (c) High-resolution spectrum of one rotational line showing isotopic structure and hyperfine structure in ¹⁹⁵Pt.

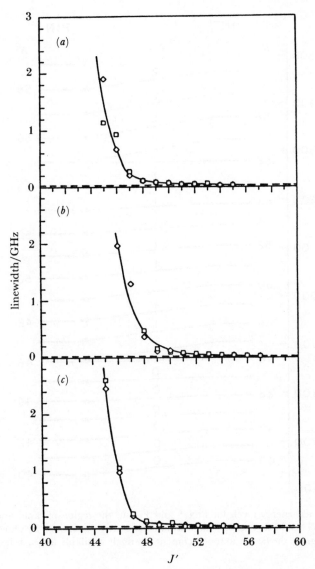

FIGURE 8. Autodetachment rate (linewidth) against rotational quantum number for various branches in PtN⁻.
(a) P-branch; (b) Q-branch, (c) R-branch.

linewidth compared to the common width observed for the corresponding P- and R-branches. This is possible because the Q-branch of a $^1\Pi$–$^1\Sigma$ transition accesses a different upper-state level than either the P- or R-branch transitions access. However, the difference in autodetachment rate for the two Λ-doublet levels in the $^1\Pi$ state is not nearly as striking as in NH⁻. (The Λ-doublet levels in PtN⁻ differ in total energy by *ca.* 0.6 cm⁻¹ at $J' = 50$.)

The need for the energy levels of the neutral PtN prompted our colleagues (D. G. Leopold, J. Ho, K. Ervin & W. C. Lineberger, unpublished results) to obtain a photoelectron spectrum of PtN⁻. This yielded an electron affinity of *ca.* 16400 cm⁻¹ and, when combined with the rotational constant B for PtN⁻, they obtained a rotational constant for the neutral PtN, $B \approx$ 0.45 cm⁻¹. From these results, Murray *et al.* (1987) were able to obtain a qualitative explanation for the autodetachment results with the help of figure 9. In the figure, a few of the rotational

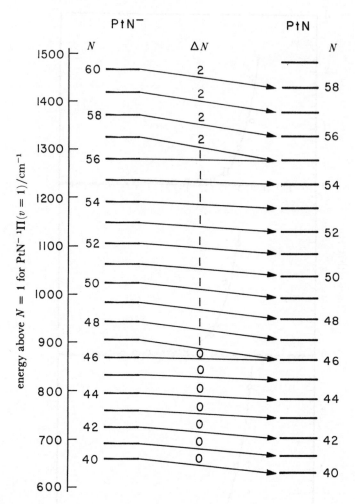

FIGURE 9. Simplified rotational energy levels for PtN⁻* and PtN in the region of rapid linewidth variation. The relation between the PtN⁻ and PtN energy scales was independently determined by photoelectron spectroscopy of PtN⁻. The arrows signify the fastest autodetachment channels available for each level.

levels for PtN are at the right and the levels for PtN⁻ are at the left. These levels are placed on the same energy scale by the photoelectron experiment, and they cannot be shifted by more than *ca.* 40 cm⁻¹ with respect to each other. The transitions for the 'best' open channels are shown in the middle. The autodetachment spectroscopic results indicated that a $v' = 1 \leftarrow v'' = 0$ transition is the one observed with a term energy of *ca.* 16445 cm⁻¹. The photoelectron results indicate an electron affinity of *ca.* 16400 cm⁻¹. Therefore, the autodetachment transitions shown in figure 9 are $\Delta v = -1$ and are extremely rapid if there is a small angular-momentum barrier.

As in all of the other systems that have been studied, autodetachment from PtN⁻ proceeds via transitions that require the consideration of certain propensity rules. We have already stated that the preferred change in vibration is $\Delta v = -1$. Now the change in nuclear rotational motion (ΔN) should be small because the low-energy, outgoing electron cannot impart much force on the nuclei. Also, the angular momentum of the outgoing electron (l) must be small to surmount the centrifugal barrier $[l(l+1)]/r^2$. The autodetachment transition rate will also increase if the electron can leave with more energy (this will go *ca.* $(eKE)^{l+\frac{1}{2}}$, where eKE is the electron kinetic energy). Taking into account these propensity rules, all of the lower Js in PtN$^-$ can detach very rapidly via

$$\Delta v = -1,$$
$$\Delta N = 0 \quad (N = J \quad \text{for PtN}^-; N = J \pm \tfrac{1}{2} \quad \text{for PtN}),$$
$$l = 0, 1 \quad \text{(i.e. s- or p-waves)}.$$

These channels will be open until $J \approx 45$, where the $\Delta N = 0$ channel shuts off. This will also increase the l required because the transition must follow the 'triangle rule'; the angular momentum of PtN$^-$ (here it is J) must equal the vector sum of the angular momentum of PtN and the angular momentum of the outgoing electron. Therefore, when the $\Delta N = 0$ channel closes at $N \approx 45$, the electron must now tunnel through a large angular-momentum barrier, and the autodetachment rate decreases. At $N \approx 57$, the $\Delta N = -1$ channel shuts off, and the $\Delta N = -2$ channel is left as the fastest transition available. This means that the electron must tunnel through an even larger barrier and the negative ion excited state lives much longer than the transit time (or fluorescence time) and will not be observed. This is very much the same case as FeO$^-$($B^4\Delta_{\frac{7}{2}}$) (Andersen *et al.* 1987), but with the channel openings and closings swapped; the transitions to higher Js broadened out.

Conclusions

The technique of high-resolution autodetachment spectroscopy has been shown to be very useful in understanding the dynamics involved in autodetachment. In this paper, only three representative systems were discussed, although at least seven systems have been studied. There is still much to be learned from negative-ion spectroscopy.

We thank our many colleagues in the study of negative ions. This work was supported by U.S. National Science Foundation grants PHY86-04504 and CHE83-16628.

References

Al-za'al, M., Miller, H. C. & Farley, J. W. 1986 *Chem. Phys. Lett.* **131**, 56.
Andersen, T., Lykke, K. R., Neumark, D. M. & Lineberger, W. C. 1986 *Electronic and atomic collisions* (ed. D. C. Lorents, W. E. Meyerhof & J. R. Peterson), pp. 791–798. Amsterdam: Elsevier.
Andersen, T., Lykke, K. R., Neumark, D. M. & Lineberger, W. C. 1987 *J. chem. Phys.* **86**, 1858.
Andresen, P. & Rothe, E. W. 1985 *J. chem. Phys.* **82**, 3634.
Berry, R. S. 1966 *J. chem. Phys.* **45**, 1228.
Crawford, O. H. 1967 *Proc. phys. Soc. London* **91**, 279.
Engelking, P. C. & Lineberger, W. C. 1976 *J. chem. Phys.* **65**, 4323.
Fano, U. 1961 *Phys. Rev.* **124**, 1866.
Gallas, J. A. C., Leuchs, G. & Walther, H. 1985 *Adv. atom. mol. Phys.* **20**, 413.
Garrett, W. R. 1970 *Chem. Phys. Lett.* **5**, 393.
Garrett, W. R. 1971 *Phys. Rev.* A **3**, 961.
Garrett, W. R. 1980 *J. chem. Phys.* **73**, 5721.
Garrett, W. R. 1982 *J. chem. Phys.* **77**, 3666.
Hefter, U., Mead, R. D., Schulz, P. A. & Lineberger, W. C. 1983 *Phys. Rev.* A **28**, 1429.
Helm, H., Cosby, P. C., Graff, M. M. & Moseley, J. T. 1982 *Phys. Rev.* A **25**, 304.
Herzberg, G. 1950 *Spectra of diatomic molecules.* New York: Van Nostrand.
Herzberg, G. 1967 *Electronic spectra of polyatomic molecules.* New York: Van Nostrand.
Herzberg, G. & Lagerqvist, A. 1968 *Can. J. Phys.* **46**, 2363.
Hotop, H. & Lineberger, W. C. 1973 *J. chem. Phys.* **58**, 2379.
Hotop, H., Patterson, T. A. & Lineberger, W. C. 1974 *J. chem. Phys.* **60**, 1806.

Jackson, R. L., Hiberty, P. C. & Brauman, J. I. 1981 *J. chem. Phys.* **74**, 3705.

Jackson, R. L., Zimmerman, A. H. & Brauman, J. I. 1979 *J. chem. Phys.* **71**, 2088.

Jones, P. L., Mead, R. D., Kohler, B. E., Rosner, S. D. & Lineberger, W. C. 1980 *J. chem. Phys.* **73**, 4419.

Kawaguchi, K. & Hirota, E. 1986 *J. chem. Phys.* **84**, 2953.

Lefebvre-Brion, H., Dehmer, P. M. & Chupka, W. A. 1986 *J. chem. Phys.* **85**, 45.

Lefebvre-Brion, H., Giusti-Suzor, A. & Raseev, G. 1985 *J. chem. Phys.* **83**, 1557.

Lineberger, W. C. & Patterson, T. A. 1972 *Chem. Phys. Lett.* **13**, 40.

Lykke, K. R., Mead, R. D. & Lineberger, W. C. 1984 *Phys. Rev. Lett.* **52**, 2221.

Lykke, K. R., Neumark, D. M., Andersen, T., Trapa, V. J. & Lineberger, W. C. 1987 *J. chem. Phys.* (In the press.)

Marks, J., Brauman, J. I., Mead, R. D., Lykke, K. R. & Lineberger, W. C. 1987 *J. chem. Phys.* (Submitted.)

Marks, J., Wetzel, D. M., Comita, P. B. & Brauman, J. I. 1986 *J. chem. Phys.* **84**, 284.

Massey, H. 1976 *Negative ions.* Cambridge University Press.

Mead, R. D., Hefter, U., Schulz, P. A. & Lineberger, W. C. 1985 *J. chem. Phys.* **82**, 1723.

Mead, R. D., Lykke, K. R., Lineberger, W. C., Marks, J. & Brauman, J. I. 1984 *J. chem. Phys.* **81**, 4883.

Moran, S., Ellis, H. B. Jr, DeFrees, D. J. & Ellison, J. B. 1987 *J. Am. chem. Soc.* (In the press.)

Murray, K. K., Lykke, K. R. & Lineberger, W. C. 1987 *Phys. Rev.* A **36**, 699.

Neumark, D. M., Lykke, K. R., Andersen, T. & Lineberger, W. C. 1985 *J. chem. Phys.* **83**, 4364.

Rehfuss, B. D., Crofton, M. W. & Oka, T. 1986 *J. chem. Phys.* **85**, 1785.

Rosenbaum, N. H., Owrutsky, J. C., Tack, L. M. & Saykally, R. J. 1986 *J. chem. Phys.* **84**, 5308.

Rosmus, P. & Meyer, W. 1978 *J. chem. Phys.* **69**, 2745.

Saykally, R. J., 1986 *Spectroscopy* **1**, 40.

Schulz, P. A., Mead, R. D., Jones, P. L. & Lineberger, W. C. 1982 *J. chem. Phys.* **77**, 1153.

Tack, L. M., Rosenbaum, N. H., Owrutsky, J. C. & Saykally, R. J. 1986 *J. chem. Phys.* **84**, 7056.

Turner, J. E. 1977 *Am. J. Phys.* **45**, 758.

Zimmerman, A. H. & Brauman, J. I. 1977a *J. chem. Phys.* **66**, 5823.

Zimmerman, A. H. & Brauman, J. I. 1977b *J. Am. chem. Soc.* **99**, 3565.

Phil. Trans. R. Soc. Lond. A **324**, 197–207 (1988)

Printed in Great Britain

Photoelectron spectroscopy of reactive intermediates

By V. Butcher, M. C. R. Cockett, J. M. Dyke, A. M. Ellis, M. Feher,
A. Morris and H. Zamanpour

Department of Chemistry, University of Southampton, Southampton SO9 5NH, U.K.

Recent developments in the use of photoelectron spectroscopy to study reactive
intermediates in the gas phase are reviewed. The information to be derived on low-
lying cationic states from such studies is illustrated by considering two diatomic
molecules, NCl and PF, and one triatomic molecule, HNO.

Also, the use of a transition-metal photoelectron spectrum to interpret the photoelectron spectrum of the corresponding transition-metal oxide is discussed by using
the spectra of vanadium and vanadium monoxide as examples. The value of superheating in high-temperature photoelectron spectroscopy is demonstrated by considering the vapour-phase photoelectron spectra of the monomers and dimers of
sodium hydroxide.

Unlike most of the other contributions to this Meeting where the aim is to prepare molecular
ions in sufficient concentration for spectroscopic study, the work described in this paper is based
on generating unstable molecules in sufficient concentrations in the gas phase for study by
photoelectron spectroscopy (PES) (Dyke *et al.* 1979, 1982*c*; Dyke 1987).

If a photon removes an electron from a molecule, then measurement of the kinetic energy
of the electrons produced combined with the conservation-of-energy condition means that the
ionization energy of the molecule can be determined. This can be summarized as

$$M + h\nu \rightarrow M^+ + e^-, \tag{1}$$

$$\text{KE}(e^-) = h\nu - I_n - \Delta E_{\text{vib}}, \tag{2}$$

where M is the molecule under consideration, ΔE_{vib} is the change in vibrational energy on
ionization and I_n is the ionization energy corresponding to the separation of the zeroth
vibrational levels in the molecule and ion. In general terms, through measurement of I_n and
ΔE_{vib} the PES technique is capable of providing information on electronic states of molecular
ions that are accessible from the neutral molecule by one-electron ionization, although if the
neutral molecule is electronically or vibrationally excited, information on the neutral species
may be obtained.

One way of generating an excited state of a molecule is via a microwave discharge of
a flowing gas mixture. For example, microwave discharging molecular oxygen excites some
$O_2(X^3\Sigma_g^-)$ to $O_2(a^1\Delta_g)$ and one-electron ionization of $O_2(a^1\Delta_g)$ gives some states of O_2^+ that are
not seen on one-electron ionization of $(O_2 X^3\Sigma_g^-)$ (Jonathan *et al.* 1974).

Another way of populating an excited state of a molecule is via an exothermic chemical
reaction. For example, NF is isoelectronic with O_2, and its first excited state, the $a^1\Delta$ state, can
be produced from the reaction of fluorine atoms with the N_3 radical, which in turn can be
produced from the reaction of fluorine atoms with HN_3 (Dyke *et al.* 1982*a, b*), i.e.

$$F + HN_3 \rightarrow N_3 + HF, \tag{3}$$

$$F + N_3 \rightarrow NF + N_2. \tag{4}$$

[123]

Both of these reactions are known to be rapid at room temperature (Pritt & Coombe 1980) and the $F + N_3$ reaction is sufficiently exothermic to populate both the $X^3\Sigma^-$ and $a^1\Delta$ states of NF.

Bands associated with ionization of both states can be seen in the experimental photoelectron spectrum recorded for the products of this reaction (see figure 1). The first ionization process

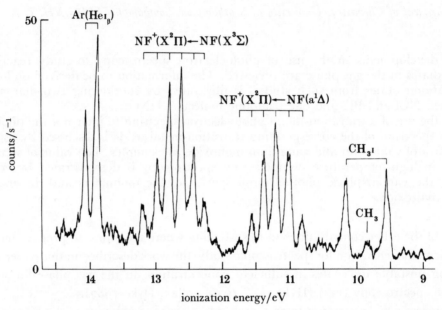

FIGURE 1. Part of the HeI photoelectron spectrum observed for the $F + N_3$ reaction.

for both the $NF(X^3\Sigma^-)$ and $NF(a^1\Delta)$ states corresponds to removal of an electron from the outermost filled orbital, a π-antibonding molecular orbital, to give the $NF^+(X^2\Pi)$ state. Measurement of the vibrational component separations in each band allows a value of the vibrational constant, $\bar{\omega}_e$, of (1520 ± 40) cm^{-1} to be determined for $NF^+(X^2\Pi)$. Also, use of the vibrational component intensities in each band, together with a Franck–Condon analysis (Dyke et al. 1979, 1982c; Dyke 1987), allows the equilibrium bond length in the ion to be determined as (1.180 ± 0.006) Å†. As expected because both the ionizations, $NF^+(X^2\Pi) \leftarrow NF(X^3\Sigma^-)$ and $NF^+(X^2\Pi) \leftarrow NF(a^1\Delta)$, correspond to removal of an electron from the outermost orbital, an antibonding π molecular orbital, the vibrational constant, $\bar{\omega}_e$, in the ion is greater than that in the neutral states ($X^3\Sigma^-$ and $a^1\Delta$) and the equilibrium bond length in the ion is lower than the neutral molecule values.

Assignment of the NF bands shown in figure 1 was achieved on the basis of a number of pieces of evidence.

(a) Both bands were clearly associated with ionization of a short-lived molecule as increasing the mixing distance above the photoionization point above 5 cm (the optimum value for producing maximum signal intensity) i.e. increasing the reaction time, produced a dramatic decrease in intensity.

(b) The separation of the first components in each band was in good agreement with the

† $1\,\text{Å} = 10^{-10}\,\text{m} = 10^{-1}\,\text{nm}$.

[124]

known separation of the zeroth vibrational levels in the $a^1\Delta$ and $X^3\Sigma^-$ neutral states, as determined from the $a^1\Delta - X^3\Sigma^-$ NF emission spectrum (Jones 1967).

(c) The experimental vertical ionization energies are in good agreement with those computed by using *ab initio* molecular-orbital calculations for each state (see table 1).

Table 1 shows ΔSCF (self-consistent field) vertical ionization energies that were obtained by performing separate SCF calculations, by using a gaussian basis set of triple-zeta plus polarization

TABLE 1. COMPUTED FIRST VERTICAL IONIZATION ENERGIES (ELECTRONVOLTS) OF $NF(X^3\Sigma^-)$
AND $NF(a^1\Delta)$

initial state	ionic state	ΔSCF value	ΔSCF + CI value	experimental value
$X^3\Sigma^-$	$^2\Pi$	13.19	12.51	12.63
$a^1\Delta$	$^2\Pi$	11.31	10.82	11.21

quality, on the neutral molecule and ion at the experimental geometry of the neutral molecule. As can be seen from this table, the ΔSCF values are higher than the experimental values indicating that unusually the correlation energy in the NF^+ $(X^2\Pi)$ state is greater than in the neutral-molecule initial state. However, in line with this interpretation, when correlation energy is allowed for from extensive configuration interaction calculations, the ΔSCF values are corrected downwards giving good agreement with experimental values. This effect has also been observed in recent independent calculations on NF and NF^+ (Bettendorff & Peyerimhoff 1985).

A similar reaction sequence could also be used to produce the NCl molecule, a molecule that is valence isoelectronic with O_2. The obvious route would be via the $Cl + HN_3$ reaction to give N_3, followed by the reaction of chlorine atoms with N_3 to give NCl. This approach was tried and found to give very little reaction, and no photoelectron bands were seen that could be associated with the NCl molecule. However, inspection of known rate constants for these reactions (Pritt & Coombe 1980) shows that the rate constant for the $Cl + HN_3$ reaction is at least one order of magnitude smaller than the rate constant for the $F + HN_3$ reaction, whereas the $Cl + N_3$ reaction has a rate constant at room temperature that is an order of magnitude larger than that for the $F + HN_3$ reaction. As a result, the following 'hybrid' reaction scheme was used to produce the NCl molecule

$$F + HN_3 \rightarrow HF + N_3, \tag{5}$$

$$Cl + N_3 \rightarrow NCl + N_2. \tag{6}$$

Part of the He I photoelectron spectrum obtained from these reactions is shown in figure 2. This shows contributions from hydrazoic acid, a reactant, and ethyl bromide, which was added as a calibrant, as well as three components of a vibrational series in the 9.5–10.5 eV ionization energy region with a vertical ionization energy of (9.82 ± 0.01) eV. This band was only observed when N_3, F atoms and Cl atoms were present and studies of the intensity of this band as a function of reaction time show clearly that it is associated with a short-lived molecule. Furthermore, analysis of the vibrational separations in the observed band gave $\omega_e = (1180 \pm 30)$ cm^{-1} in the ionic state compared with values of $\bar\omega_e$ in NCl $X^3\Sigma^-$ and NCl $a^1\Delta$ of 827 and 905 cm^{-1} respectively (Huber & Herzberg 1979; Pritt *et al.* 1981). The observed value of $\bar\omega_e$ in the ion is therefore greater than the NCl neutral values, consistent with ionization from the outermost molecular orbital, which is antibonding in character.

[125]

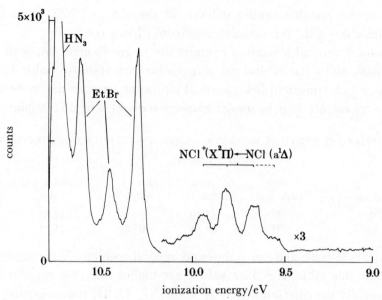

FIGURE 2. Part of the He I photoelectron spectrum observed for the $Cl + N_3$ reaction.

Although these observations are consistent with assignment of the band at 9.82 eV to ionization of NCl, unlike the NF case only one band was observed despite the fact that the $Cl + N_3$ reaction is sufficiently exothermic to populate the $^1\Delta$ and $^3\Sigma^-$ states (Clyne & MacRobert 1983). One obvious interpretation is that the observed band is caused by the $NCl^+(X^2\Pi) \leftarrow NCl(a^1\Delta)$ ionization with the corresponding $NCl^+(X^2\Pi) \leftarrow NCl(X^3\Sigma^-)$ band being masked by more intense bands in the 10.5–11.5 eV region.

To test this hypothesis, *ab initio* calculations that include the effects of electron correlation were performed for NCl and NCl^+ by using the approach adopted previously for NF and NF^+. The results of these calculations are shown in table 2. Again, as in the NF case, the ΔSCF vertical ionization energy for the $NCl^+(X^2\Pi) \leftarrow NCl(a^1\Delta)$ ionization is too high, suggesting that the correlation energy in the ion is greater than that in the neutral molecule. However, allowing for correlation energy in each state via extensive configuration interaction calculations brings the computed $NCl^+(X^2\Pi) \leftarrow NCl(a^1\Delta)$ vertical ionization energy in good agreement with experiment.

TABLE 2. COMPUTED FIRST VERTICAL IONIZATION ENERGIES (ELECTRONVOLTS) OF $NCl(X^3\Sigma^-)$
AND $NCl(a^1\Delta)$

initial state	ionic state	ΔSCF value	ΔSCF + CI value	experimental value
$X^3\Sigma^-$	$^2\Pi$	12.08	11.33	—
$a^1\Delta$	$^2\Pi$	10.33	9.84	9.82

The calculated vertical ionization energies shown in table 2 support the hypothesis that the $NCl^+(X^2\Pi) \leftarrow NCl(a^1\Delta)$ ionization has been observed and that the $NCl^+(X^2\Pi) \leftarrow NCl(X^3\Sigma^-)$ ionization is not seen because it is overlapped by the more intense bands in the 10.5–11.5 eV region.

As in the NF case, the vibrational component intensities in the observed NCl band can be used with a series of Franck–Condon factor calculations to determine the decrease in equilibrium bond length on ionization as (0.09 ± 0.01) Å. Unfortunately, it appears that the equilibrium bond length in NCl $(a^1\Delta)$ has not been determined experimentally and as a result it is not possible to calculate the equilibrium bond length, r_e, in $NCl^+(X^2\Pi)$.

The low ionization-energy part of the photoelectron spectrum of PF, a molecule that is valence isoelectronic with oxygen, is also similar to that observed for NCl. For PF, the neutral molecule has been produced as a secondary product of the $F + PH_3$ reaction and the first band attributable to ionization of this molecule at a vertical ionization energy of (9.74 ± 0.01) eV (Dyke *et al.* 1982*c*; see figure 3) is assigned on the basis of *ab initio* calculations that include the

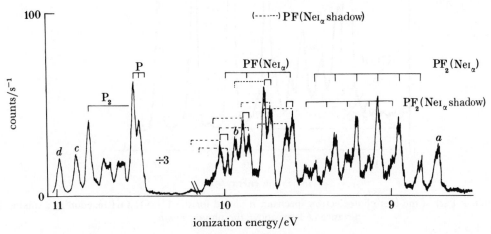

FIGURE 3. The Ne I photoelectron spectrum of the first band of PF. Bands labelled *a, c* and *d* are obtained by ionization of helium. The PF is produced as a secondary product of the $F + PH_3$ reaction. Bands: *a*, He ionized by Ne III (32.68 eV); *b*, PF (Ne I$_\alpha$ shadow of second component; *c*, He ionized by Ne II (30.54 eV); *d*, He ionized by Ne II (30.45 eV).

effects of electron correlation to the $PF^+(X^2\Pi) \leftarrow PF(X^3\Sigma^-)$ ionization. Again, as in the NF and NCl examples, the vibrational constant, $\bar{\omega}_e$, determined for the ionic state from the experimental vibrational component separations is greater then that in the neutral molecule. The measured value in $PF^+(X^2\Pi)$ is (1030 ± 30) cm^{-1} compared with the corresponding value in the $PF(X^3\Sigma^-)$ state of 847 cm^{-1} (Huber & Herzberg 1979), and the use of the vibrational component intensities via a series of Franck–Condon calculations allows the equilibrium bond length in the ion to be determined as (1.498 ± 0.005) Å. Also as shown in figure 3, each vibrational component in the $PF^+(X^2\Pi) \leftarrow PF(X^3\Sigma^-)$ band is resolved into two parts. This arises from spin–orbit splitting in the $PF^+(X^2\Pi)$ state and the measured average value for this separation, (316 ± 16) cm^{-1}, is in good agreement with a value of 324 cm^{-1} determined in a study of the $PF^+(A^2\Sigma^+) \rightarrow PF^+(X^2\Pi)$ emission spectrum (Douglas & Frackowiak 1966).

HNO is a triatomic molecule that is isoelectronic with O_2. It can be conveniently prepared in the gas phase by the reaction of fluorine atoms with hydroxylamine, NH_2OH, and part of the photoelectron spectrum obtained from this reaction recorded at a reagent-mixing distance of 1 cm above the photon beam is shown in figure 4. As well as some residual hydroxylamine, this spectrum shows the first band of nitric oxide, a reaction product, as well as two other bands that were also found to be associated with reaction products; a sharp intense band at (10.77 ± 0.01) eV and a broad, structured series with a vertical ionization energy of

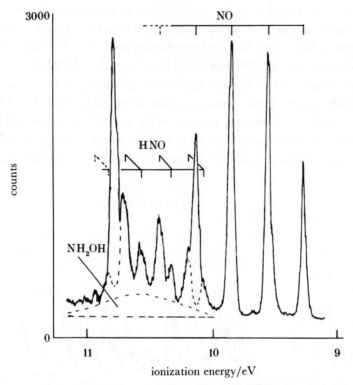

FIGURE 4. Part of the HeI photoelectron spectrum recorded for the $F + NH_2OH$ reaction at a mixing distance of 1 cm above the photon beam.

(10.68 ± 0.01) eV. When the intensities of these bands were investigated as a function of reaction time, both bands were found to be associated with reactive intermediates, but the band at 10.77 eV was found to be associated with a much shorter-lived molecule than the band at 10.68 eV. On the basis of this evidence the broad band centred at 10.68 eV is assigned to the first ionization of HNO and the band at 10.77 eV is tentatively assigned to ionization of HNOH. The assignment of the 10.68 eV band to HNO is a lot firmer than the assignment of the band at 10.77 eV as the adiabatic ionization energy of the 10.68 eV band, measured as (10.07 ± 0.03) eV, is in reasonably good agreement with the first adiabatic ionization energy of HNO of (10.29 ± 0.14) eV determined by electron-impact mass spectrometry (Kohout & Lampe 1966). Also, the observed vibrational structure in this band can be assigned to excitation of the N–O stretching mode and the HNO deformation mode in the ion; the N–O stretching frequency is increased over the corresponding value in the neutral molecule, $HNO(X^1A')$, as expected by analogy with oxygen, as ionization occurs from a molecular orbital that is antibonding in the N–O direction, whereas the deformation frequency is decreased from the value in $HNO(X^1A')$. The measured average separations are (1940 ± 40) cm^{-1} for the N–O stretch and (880 ± 40) cm^{-1} for the deformation mode whereas the corresponding values in $HNO(X^1A')$ are 1556 cm^{-1} (ν_2) and 1501 cm^{-1} (ν_3) (Johns et al. 1983). These changes in vibrational frequencies on ionization are consistent with the equilibrium geometry change between $HNO(X^1A')$ and $HNO^+(X^2A')$ expected from *ab initio* molecular-orbital calculations (Bruna 1980; Bruna & Marian 1979).

As well as investigating reactive intermediates by single-photon ionization, it is also possible to investigate molecules of this type by multiphoton ionization. By recording the total ion–current as a function of laser wavelength it is possible to probe the spectroscopic properties of resonant intermediate states whereas by recording a photoelectron spectrum at a fixed laser wavelength, it is possible to probe the spectroscopic properties of the ion. A very simple spectrometer has been constructed with a borrowed excimer pumped dye laser to demonstrate that short-lived molecules can be investigated in this way.

An MPI (multi-photon ionization) ion–current spectrum recorded for discharged oxygen, and tentatively assigned to $O_2(^1\Delta_g)$, has been recorded in the wavelength range 455.0–430.0 nm. The linewidth of the spectrum was controlled by the Doppler effect and the laser linewidth. Although photoelectron spectra at selected laser wavelengths need to be recorded and the laser power dependence of the observed signals needs to be investigated, the experimental ion spectrum is currently being analysed in terms of a $(3+1)$ ionization process via a number of as-yet unobserved singlet Rydberg states of oxygen. The use of this study lies in the fact that MPI studies on $O_2(X^3\Sigma_g^-)$ have the ability to probe excited triplet states accessible from the ground $^3\Sigma_g^-$ state (Katsumata *et al.* 1986), whereas similar studies on $O_2(^1\Delta_g)$ have the advantage in that singlet excited states can be probed, for which much less experimental information is currently available. More generally, this approach is valuable in that photoelectron spectra can be recorded at selected laser wavelengths and, as a result, simplified vibrational structure will usually be seen compared with that observed in a single photon spectrum. This arises because highly excited states of neutrals often have similar equilibrium geometries and spectroscopic constants to the ground state of the positive ion. Hence short vibrational series will usually be seen in the experimental photoelectron spectra. By preparing different vibronic excited intermediate states, different vibrational frequencies will be excited in the ion. Hence in a polyatomic free radical such as the phenyl radical, where the structure in the single photon photoelectron spectrum is very complex (Butcher *et al.* 1987), it should be possible to observe considerably simplified structure. It should also be possible to determine the fundamental frequencies of a number of modes of the ion simply by preparing different highly excited vibronic states for ionization.

As well as having interests in reactive intermediates produced in the gas phase by microwave discharge or rapid atom–molecule reactions, we also have interests in molecules produced by high-temperature evaporation. Evaporation temperatures of up to 2800 K have been achieved by using a radiofrequency induction heating method (Morris *et al.* 1986). The main areas of investigation have centred on the study of metals, metal oxides and metal hydroxides.

Recently, the HeI photoelectron spectra of all the first-row transition metals has been studied (Dyke *et al.* 1985 a). These elements were investigated firstly as possible precursors of transition-metal oxides in the vapour phase and secondly to obtain the 4s:3d photoionization cross sections in these elements to provide reliable data to be used in the interpretation of intensities of bands recorded in the photoelectron spectra of transition-metal compounds, notably metal oxides. It has been found that the 4s:3d cross-section ratio at the HeI photon energy is very low at scandium and titanium, but it increases fairly regularly to copper and zinc. As an example of a typical spectrum, figure 5 shows the HeI photoelectron spectrum of vanadium (Dyke *et al.* 1985 a).

In this diagram, bands B and C are $(4s)^{-1}$ ionizations of the ground state of atomic vanadium whereas band E is a $(3d)^{-1}$ ionization of the ground state. Measurement of the relative band intensities allows the 4s:3d photoionization cross-section ratio of vanadium to be determined

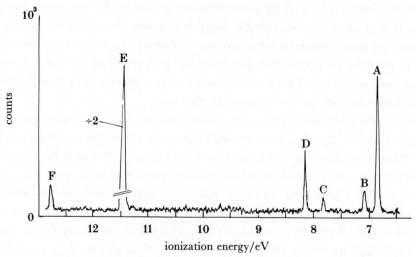

FIGURE 5. The HeI photoelectron spectrum of atomic vanadium.

at the HeI photon energy as $1:(29.8 \pm 2.5)$, indicating that the 3d cross section is much greater than the 4s cross section. Bands A and D in figure 5 cannot be assigned to ground-state ionizations and are $(3d)^{-1}$ ionizations of an excited state of vanadium, approximately 2200 cm^{-1} above the ground state. The relative A:B band-intensity ratio can be used with the previously calculated 4s:3d cross-section ratio to give an effective beam temperature at the point of photoionization of 1940 K, which is slightly less than the furnace temperature used for evaporation of the metal of (2030 ± 30) K. This is typical of all the transition metals studied where the beam temperature at the point of photoionization, as evaluated from atomic band intensities, is several hundred degrees below the furnace temperature.

Metal oxides can also be generated in the vapour phase by direct evaporation and knowledge

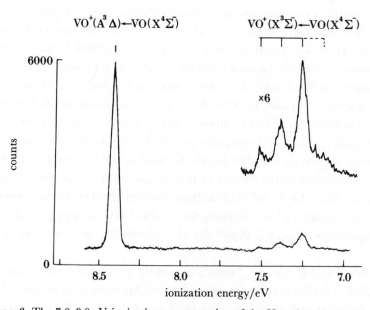

FIGURE 6. The 7.0–9.0 eV ionization energy region of the HeI photoelectron spectrum of VO.

of the photoelectron spectrum of the appropriate transition metal is very important in assigning the experimental spectrum. For example, vanadium monoxide can be obtained in the vapour phase by vaporizing stoichiometric vanadium monoxide (VO(s)) from a tungsten furnace at 2000 K (Dyke *et al.* 1985 *b*). The spectrum obtained in the 7.5–8.5 eV ionization energy region is shown in figure 6. The first band of VO, shown in this figure, is essentially a metal $(4s)^{-1}$ ionization whereas the second band is essentially a metal $(3d)^{-1}$ ionization, and this explains qualitatively why the first band is much weaker than the second. This assignment has been derived from the results of Hartree–Fock–Slater calculations and *ab initio* molecular-orbital calculations performed on the ground state of VO, the $X^4\Sigma^-$ state, and the two lowest-lying ionic states obtained by one-electron ionization from the neutral molecule. As can be seen from figure 6, the first band shows clear vibrational structure whereas the second band shows only one vibrational component. Measurement of the vibrational separations in the first band of VO allows the vibrational constant, $\bar{\omega}_e$, to be determined as (1060 ± 40) cm^{-1} in the ground state of VO$^+$, the $X^3\Sigma^-$ state, and use of the vibrational component intensities, via a series of Franck–Condon calculations, allows the equilibrium bond length in the ion to be determined as (1.54 ± 0.01) Å.

As an example of our recent interest in metal hydroxides, the photoelectron spectra of sodium hydroxide will be briefly discussed (Dyke *et al.* 1986). For this hydroxide, mass-spectrometric studies have shown that in the temperature range 600–700 K, dimers are a major vapour-phase constituent and, in view of this, superheating beams of this hydroxide at temperatures up to 300 K higher than the furnace temperature has been used with the aim of simplifying the spectra obtained and obtaining the photoelectron spectrum of the monomer.

The He I photoelectron spectrum obtained for sodium hydroxide heated in a silver-lined carbon furnace at (720 ± 50) K is shown in figure 7 *a* and the spectrum obtained with super-

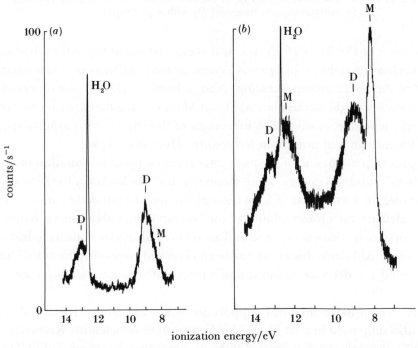

FIGURE 7. The He I photoelectron spectrum of NaOH recorded (*a*) without superheating and (*b*) with superheating.

heating is shown in figure 7b. On the basis of these spectra, bands marked M are assigned to ionizations of the monomer whereas the bands marked D are assigned to ionization of the dimer, as superheating is expected to increase the relative partial pressure of the monomer relative to the dimer. Also, from the results of *ab initio* molecular-orbital calculations the two observed monomer bands are assigned to ionization from the outermost π and σ monomer molecular orbitals.

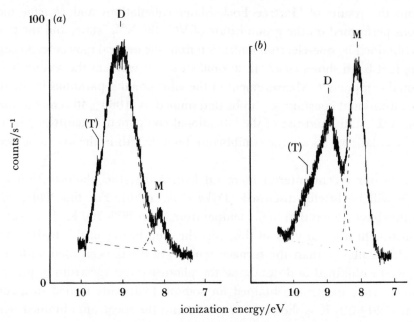

FIGURE 8. The 7.0–11.0 eV ionization-energy region of the HeI photoelectron spectrum of NaOH recorded (a) without superheating and (b) with superheating.

An expanded view of the 7.0–10.0 eV spectral region recorded for sodium hydroxide with and without superheating is shown in figure 8. As can be seen, the monomer first band increases relative to that of the dimer on superheating. Also, a band marked T was observed that was found to be independent of the bands marked D and M on varying the experimental conditions, and this band was tentatively assigned to ionization of the trimer. Very similar spectra have been obtained for lithium and potassium hydroxide (Dyke *et al.* 1986).

From these spectra, perhaps one of the most important pieces of information to be obtained is the first adiabatic ionization energy of the monomer and this leads via the heat of formation of the neutral molecule, to the heat of formation of the metal-hydroxide cation. This in turn can be used to calculate the proton affinity of the corresponding alkali-metal oxide. It is also notable that the first ionization energy of the alkali-metal hydroxides appears to increase as the cluster size increases. Although this effect has been observed previously for alkali halides, the more common trend is a decrease in ionization energy with increasing cluster size.

We gratefully acknowledge financial support for this work from SERC and the CEGB. This work was also supported in part by the Air Force Office of Scientific Research (grant no. AFOSR-83-0283) through the European Office of Aerospace Research (EOARD), United States Air Force.

References

Bettendorff, M. & Peyerimhoff, S. D. 1985 *Chem. Phys.* **99**, 55.

Bruna, P. J. 1980 *Chem. Phys.* **49**, 39.

Bruna, P. J. & Marian, C. M. 1979 *Chem. Phys.* **37**, 425.

Butcher, V., Costa, M. L., Dyke, J. M., Ellis, A. R. & Morris, A. 1987 *Chem. Phys.* **115**, 261.

Clyne, M. A. A. & MacRobert, A. J. 1983 *J. chem. Soc. Faraday Trans.* II **79**, 283.

Douglas, A. E. & Frackowiak, M. 1966 *Can. J. Phys.* **45**, 1074.

Dyke, J. M. 1987 *J. chem. Soc. Faraday Trans.* II **83**, 69.

Dyke, J. M., Feher, M. & Morris, A. 1986 *J. Electron. Spectrosc. rel. Phen.* **41**, 343.

Dyke, J. M., Gravenor, B. W. J., Hastings, M. P., Josland, G. D. & Morris, A. 1985*a* *J. Electron. Spectrosc. rel. Phen.* **35**, 65.

Dyke, J. M., Gravenor, B. W. J., Hastings, M. P. & Morris, A. 1985*b* *J. phys. Chem.* **89**, 4613.

Dyke, J. M., Lewis, A. E., Jonathan, N. & Morris, A. 1982*a* *J. chem. Soc. Faraday Trans.* II **78**, 1445.

Dyke, J. M., Lewis, A. E., Jonathan, N. & Morris, A. 1982*b* *Molec. Phys.* **47**, 1231.

Dyke, J. M., Morris, A. & Jonathan, N. 1979 *Electron spectroscopy: theory, techniques and applications*, vol. 3, p. 189. New York: Academic Press.

Dyke, J. M., Morris, A. & Jonathan, N. 1982*c* *Int. Rev. phys. Chem.* **2**, 3.

Huber, K. P. & Herzberg, G. 1979 *Molecular spectra and molecular structure. IV. Constants of diatomic molecules.* New York: Van Nostrand.

Jonathan, N., Morris, A., Okuda, M., Ross, K. J. & Smith, D. J. 1974 *J. chem. Soc. Faraday Trans.* II **70**, 1810.

Jones, W. E. 1967 *Can. J. Phys.* **45**, 21.

Johns, J. W. C., McKellar, A. R. W. & Weinberger, E. 1983 *Can. J. Phys.* **61**, 1106.

Katsumata, S., Sato, K., Achiba, Y. & Kimura, K. 1986 *J. Electron. Spectrosc. rel. Phen.* **41**, 325.

Kohout, F. C. & Lampe, F. W. 1966 *J. chem. Phys.* **45**, 1074.

Morris, A., Dyke, J. M., Josland, G. D., Hastings, M. P. & Francis, P. D. 1986 *High Temp. Sci.* **22**, 95.

Pritt, A. T. & Coombe, R. D. 1980 *Int. J. chem. Kinet.* **12**, 741.

Pritt, A. T., Patel, D. & Coombe, R. D. 1981 *J. molec. Spectrosc.* **87**, 401.

References

Phil. Trans. R. Soc. Lond. A **324**, 209–221 (1988)

Printed in Great Britain

Rotational and vibronic structure in the electronic spectra of linear open-shell cations

By J. P. Maier

Institut für Physikalische Chemie, Universität Basel, Klingelbergstrasse 80,
CH-4056 Basel, Switzerland

The vibrational and rotational structure of the electronic transitions of open-shell cations of linear polyatomics has been studied by three complementary techniques. In one, the emission spectra of ions of unstable species, XCP^+ and XBS^+, with $X = H$, D and F, are obtained by electron-impact excitation on effusive and supersonic free jets. The vibronic structure of their $\tilde{A}^2\Sigma^+ \to \tilde{X}^2\Pi_i$ band systems could be analysed. The changes and distinct features of the rotational profiles at low rotational temperatures in the emission spectra of diacetylene cation have been used to establish the absolute numbering in the rotational analysis of the $\tilde{A}^2\Pi_u–\tilde{X}^2\Pi_g$ transition. The second approach is based on the measurement of the laser excitation spectra of ions and has been used to confirm the vibrational assignment of complex emission spectra, such as the $\tilde{B}^2\Pi \to \tilde{X}^2\Pi$ transition of $ClCN^+$, and to resolve the rotational structure in the $\tilde{A}^2\Pi_{\frac{3}{2}} \leftarrow \tilde{X}^2\Pi_{\frac{3}{2}}$ transitions of the haloacetylene cations $XCCH^+$, $X = Cl$, Br, I. The most recent development is the demonstration that the stimulated emission pumping approach can be used to probe the electronic structure of ions in a flow system.

1. Introduction

Electronic spectroscopy has a long-standing tradition of providing information on the structure and spectroscopic parameters for transient species (Herzberg 1971). Many radicals and diatomic ions were thus characterized, initially by using classical absorption and emission approaches, and subsequently further and finer details have been discovered by laser techniques (Herzberg 1985). As far as polyatomic cations are concerned, the application of laser-based methods is even more recent (Miller & Bondybey 1982). However, following the identification of the emission spectra of over a hundred open-shell organic cations in the gas phase generated by electron-beam excitation (Maier 1980), higher resolution and state-selected investigations could be undertaken (Maier 1982).

Our research interests in the last years have focused on the spectroscopic characterization of polyatomic open-shell cations by emission spectroscopy with supersonic free jets, by laser excitation of the fluorescence and absorption measurements in neon matrices (Maier 1986). The present article gives examples of some of the recently completed studies with the first two techniques to obtain new vibronic data on triatomic cations of unstable molecules, to resolve the rotational structure in the electronic transitions of substituted acetylene cations and to use the techniques hand-in-hand to provide unambiguous vibronic and rotational assignments. Additionally, the stimulated emission pumping technique has also been demonstrated to be a viable approach to probe the ground state of ions.

2. Emission spectroscopy

Electron-impact excitation of the emission spectra has proved to be a valuable method in the studies of polyatomic cations in the gas phase, especially in obtaining the first set of data on their electronic transitions (Maier 1980). The incorporation of a seeded helium supersonic free jet as the source led to further progress in this area (Klapstein *et al.* 1983). This approach was consequently chosen to provide the initial information on the optical transitions of ions of unstable species, i.e. XCP^+, X = H, D, F and XBS^+, X = H, D, F, Cl. The molecular species themselves were produced by fast-flow methods from the appropriate precursors as described in their studies by microwave and infrared (IR) spectroscopy, and have also all been subjected to photoelectron spectroscopic investigations (Kroto 1982). The latter provided the first information on their ions, albeit at low resolution. Nevertheless, these data were important in assigning the observed emission-bands systems to the electronic transitions of such cations.

The $\tilde{A}^2\Sigma^+ \to \tilde{X}^2\Pi_i$ emission-band systems of phosphorus substituted alkyne ions XCP^+, with X = H, D (King *et al.* 1981), F (King *et al.* 1984*a*), and of the related thioborines XBS^+, with X = H, D (King *et al.* 1985*a*), F and Cl (King *et al.* 1986), have thus been obtained by electron-impact excitation. Following the preliminary vibrational analyses, and for HCP^+ and DCP^+ also rotational analyses (King *et al.* 1982), the most recent studies have been focused on understanding the complex vibronic pattern in the spectra of HCP^+ and DCP^+ (King *et al.* 1987). To this end, the emission spectra have been recorded with improved signal:noise ratios in the relevant regions, and especially helpful have been the changes in the rotational profiles at low rotational temperatures attained by means of a seeded helium supersonic free jet. This is especially the case for the weaker bands for which the resolution of the rotational structure proved not to be possible. In addition, the drastic narrowing of the bands led to the detection of weak transitions (which are often obscured at higher temperatures), such as those associated with single and double quantum excitation of the ground-state ($\tilde{X}^2\Pi_i$) bending mode, v_2''.

As an example, the spectral region near the origin bands of the $\tilde{A}^2\Sigma^+ \to \tilde{X}^2\Pi_i$ transition of HCP^+ is shown in figure 1 with rotational temperatures of *ca.* 300 K (upper) and 20–30 K (bottom trace). The latter temperature is inferred by simulation of the profiles from the known rotational constants (King *et al.* 1982). The two spin–orbit components are seen ($A'' \sim -150$ cm^{-1}) and at *ca.* 300 K the P-head as well as the stronger Q-branch head are distinct. On cooling, however, the P-head disappears. Thus a comparison of two such spectra enables the high and low J transitions to be distinguished. Furthermore, the R-branches are more intense than the P-branches for the $\Omega = \frac{1}{2}$ component bands but equally strong for $\Omega = \frac{3}{2}$ ones. All these features have been used as a diagnostic tool in the assignments.

The excitation of the degenerate v_2 mode is identified by means of sequence transitions, e.g. 010–010, as well as even quanta progressions, e.g. 000–020. The discernible band systems are complicated by a combination of Renner and Fermi interactions (see figure 2). An analysis has proved possible in terms of the parameters used to describe such phenomena e.g. $\epsilon\omega_2'' \sim -26.5$ (-18.7) cm^{-1}, $g_\kappa'' = 6.4$ (3.6) cm^{-1} for the Renner effect in HCP^+ (DCP^+). Though some discrepancies remain, the essential spectral features are mimicked (King *et al.* 1987).

Vibronic analyses to such detail have also been completed for the $\tilde{A}^2\Sigma^+ \to \tilde{X}^2\Pi_i$ transitions of HBS^+ and DBS^+, and to a more limited extent for FCP^+ and recently FBS^+ and $ClBS^+$. In table 1 are summarized the pertinent spectroscopic data deduced for the XCP^+ and XBS^+ ions. In view of this information, it should in due course be feasible to obtain further and more precise data by methods based on laser excitation.

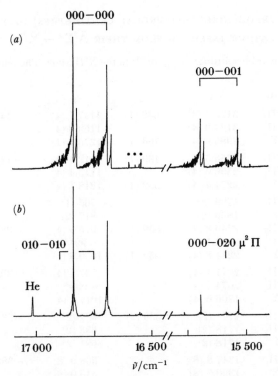

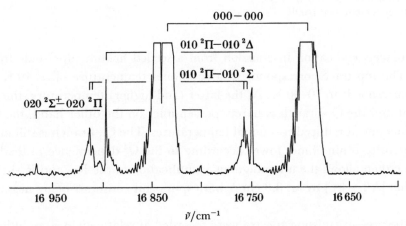

FIGURE 1. Two spectral regions in the $\tilde{A}^2\Sigma^+ \to \tilde{X}^2\Pi_i$ emission system of HCP⁺ recorded (0.065 nm FWHM) with electron impact excitation on an effusive beam of pure HCP (top) and on a seeded helium supersonic free jet (bottom).

FIGURE 2. Expanded scan of the spectral features adjoining the origin bands in the $\tilde{A}^2\Sigma^+ \to \tilde{X}^2\Pi_i$ emission spectrum of supersonically cooled HCP⁺.

For the $\tilde{A}^2\Pi_u \to \tilde{X}^2\Pi_g$ transition of the diacetylene cation, the pronounced changes in the band profiles on cooling enable the absolute numbering in the rotational analysis to be established. Although the emission spectrum was identified and rotationally analysed a long time ago (Callomon 1956), in fact two physically reasonable numberings were obtained (and hence two sets of spectroscopic constants). This ambiguity was resolved by the following procedure.

In figure 3 are shown two recordings of the origin band of the $\tilde{A}^2\Pi_{u,\Omega} \to \tilde{X}^2\Pi_{g,\Omega}$ ($\Omega = \frac{3}{2}, \frac{1}{2}$)

TABLE 1. VIBRATIONAL FREQUENCIES (RECIPROCAL CENTIMETRES) OF THE PHOSPHAETHYNE AND SULPHIDOBORON CATIONS INFERRED FROM THEIR $\tilde{A}^2\Sigma^+ \to \tilde{X}^2\Pi_i$ EMISSION SPECTRA

(Also are given the apparent spin–orbit splittings, $A_{0,\mathrm{eff}}(\mathrm{cm}^{-1})$ in the $\tilde{X}^2\Pi_i$ state. The references to the studies are to be found in the text.)

ion	state	ν_1	ν_2	ν_3	$A_{0,\mathrm{eff}}$
HCP[+a]	$\tilde{X}^2\Pi_{\frac{3}{2}}$	3125.1 (4)	642 (1)	1147.1 (4)	-146.97 (3)
	$^2\Pi_{\frac{1}{2}}$	3124.9 (4)		1159.9 (4)	
	$\tilde{A}^2\Sigma^+$	2985.6 (4)	706 (1)	1275.4 (4)	
DCP[+b]	$\tilde{X}^2\Pi_{\frac{3}{2}}$	2356.5 (4)	499 (1)	1112.4 (4)	-146.71 (1)
	$^2\Pi_{\frac{1}{2}}$	2356.6 (4)		1113.4 (4)	
	$\tilde{A}^2\Sigma^+$	2274.4 (4)	552 (1)	1218.1 (4)	
FCP$^+$	$\tilde{X}^2\Pi_i$	1729 (2)		765 (1)	-190.2 (6)
	$\tilde{A}^2\Sigma^+$	1866 (2)		817 (2)	
H^{11}B^{32}S[+c]	$\tilde{X}^2\Pi_{\frac{3}{2}}$	2746.8 (4)	659 (1)	975.9 (4)	-322.6 (4)
	$^2\Pi_{\frac{1}{2}}$	2747.1 (4)		$\sim 991^d$	
	$\tilde{A}^2\Sigma^+$	2214.8 (4)	550 (1)	1050.9 (4)	
D^{11}B^{32}S$^+$	$\tilde{X}^2\Pi_{\frac{3}{2}}$	2071.1 (4)		937.4 (4)	-322.2 (4)
	$^2\Pi_{\frac{1}{2}}$	2074.2 (4)		$\sim 993^d$	
	$\tilde{A}^2\Sigma^+$	1706.6 (4)		1011.1 (4)	
F^{11}B^{32}S$^+$	$\tilde{X}^2\Pi_{\frac{3}{2}}$	1721 (2)	339 (2)	637 (2)	-339 (2)
	$^2\Pi_{\frac{1}{2}}$	1718 (2)		633 (2)	
	$\tilde{A}^2\Sigma^+$	1718 (2)		691 (2)	
^{35}Cl^{11}B^{32}S$^+$	$\tilde{X}^2\Pi_i$	1347.8 (8)		508.9 (8)	-383 (1)
	$\tilde{A}^2\Sigma^+$	1390.6 (8)		516.0 (8)	

[a] $B_0(\tilde{X}^2\Pi_i) = 0.6224$ (16); $B_0(\tilde{A}^2\Sigma^+) = 0.6690$ (17) cm^{-1}.

[b] $B_0(\tilde{X}^2\Pi_i) = 0.5284$ (2); $B_0(\tilde{A}^2\Sigma^+) = 0.5682$ (2) cm^{-1}.

[c] $B_0(\tilde{X}^2\Pi_i) = 0.5760$ (2); $B_0(\tilde{A}^2\Sigma^+) = 0.6148$ (2) cm^{-1}. These values are taken from McDonald & Innes (1969), where they are erroneously attributed to BS.

[d] Estimated from combination bands.

transition of diacetylene cation in emission from a seeded helium supersonic free jet (Kuhn et al. 1986). The top trace corresponds to a rotational temperature of ca. 10 K whereas the bottom one corresponds to 70–90 K. At the latter (and higher) temperature, the two R-heads are prominent and the Q_1-branch is barely perceptible; on the other hand, the Q_1-branch is the most distinct one at reduced rotational temperatures. The Q_2-branch is still not discernible because its intensity is nine times lower according to the Ω^2-dependence in the line strength. This difference then allows the unambiguous identification of the R_1- and R_2-heads, showing in this case that $|A''| > |A'|$ by ca. 3.3 cm^{-1}, both spin-orbit constants being negative (inverted states).

Although the present emission spectra were recorded at relatively low resolution (0.25 cm^{-1} FWHM), the position of the Q_1-branch maximum is located to within ± 0.05 cm^{-1} and establishes the correct numbering. From the two possible sets of spectroscopic constants derived from the rotational analysis the position of the $Q_1(J = 1.5)$ line is calculated. Whereas one of the values agrees exactly with the observation (see the middle trace of figure 3), also in the case of the corresponding data and spectrum of dideuterodiacetylene cation, the other value lies 0.3 cm^{-1} to higher energy. The correct numbering is thus established. The spectroscopic constants were also newly determined for both diacetylene and dideuterodiacetylene cations by fitting simultaneously the line positions in their $\tilde{A}^2\Pi_u \leftarrow \tilde{X}^2\Pi_g$ laser excitation spectra to the eigenvalues of the usual hamiltonian matrix. The derived B_0'', B_0' values are included in the compilation of table 2.

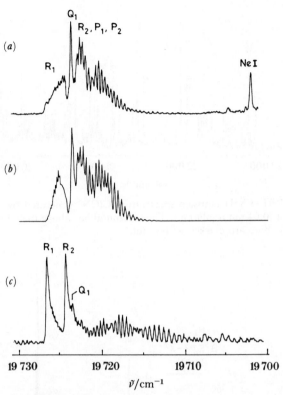

FIGURE 3. The origin band in the $\tilde{A}^2\Pi_u \rightarrow \tilde{X}^2\Pi_g$ emission spectrum of diacetylene cation obtained at 0.25 cm^{-1} resolution with electron impact on a supersonic expansion of diacetylene seeded in helium. Traces (a) and (c) correspond to rotational temperatures of 10 K and 70–90 K respectively. The middle trace (b) is a computer simulation of the transition (see text), $T = 10$ K.

3. LASER EXCITATION SPECTROSCOPY

The advantage of using emission and laser excitation spectroscopies in tandem for the spectral characterization of open-shell cations is illustrated by two examples. In the first, the vibrational analysis of the laser excitation spectrum of chlorocyanide cation was needed to locate the origin of the transition, which in turn enabled the emission spectrum to be interpreted. In the second, the initial information provided by the emission data on the haloacetylene cations led to the study of these species at higher resolution by laser excitation, and subsequently to the IR rotational and related spectroscopic constants.

In figure 4 is shown the main part of the $\tilde{B}^2\Pi_i \rightarrow \tilde{X}^2\Pi_i$ emission spectrum of rotationally cooled ($T_{\mathrm{rot}} = 5$–10 K) chlorocyanide cation excited in a seeded supersonic free jet (Fulara et al. 1985). Although the spectrum is much improved compared to that with an effusive source (Allan & Maier 1976), several vibronic assignments of the band system appear to be possible. The complex vibration pattern arises from the large geometry change taking place on passing from the $\tilde{X}^2\Pi$ to the $\tilde{B}^2\Pi$ ionic state (mainly an increase in the C–Cl distance) as well as from the strong overlap of the $^2\Pi_{\frac{3}{2}} \rightarrow {}^2\Pi_{\frac{3}{2}}$ and $^2\Pi_{\frac{1}{2}} \rightarrow {}^2\Pi_{\frac{1}{2}}$ subsystems. However, the location of their origins is readily established by recording the laser excitation spectrum of this transition (Celii et al. 1986a). In the latter approach, the ions are prepared essentially all in the lowest vibrational level of the $\tilde{X}^2\Pi_i$ state by means of Penning ionization and collisional relaxation. Consequently the intensity of the origin band (of the $\Omega = \frac{3}{2}$ subsystem) becomes appreciable in

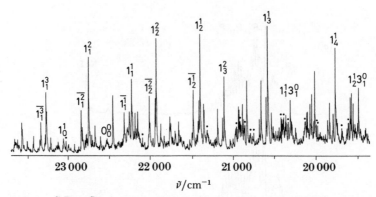

FIGURE 4. A portion of the $\tilde{B}^2\Pi \rightarrow \tilde{X}^2\Pi$ emission spectrum of ClCN$^+$ generated by electron impact on a seeded helium supersonic free jet (0.04 nm resolution). The horizontal bar above the vibrational assignment identifies the $\Omega = \frac{1}{2}$ system; atomic lines are marked with a dot.

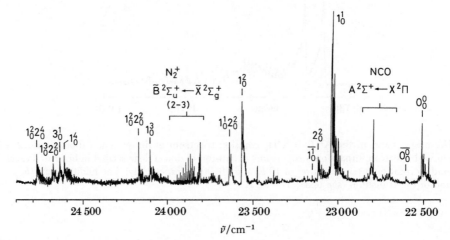

FIGURE 5. The main part of the $\tilde{B}^2\Pi \leftarrow \tilde{X}^2\Pi$ laser excitation spectrum of ClCN$^+$ recorded with a resolution of 0.2 cm^{-1}.

spite of the Franck–Condon factors (see figure 5). The origin band of the $\tilde{B}^2\Pi_{\frac{1}{2}} \leftarrow \tilde{X}^2\Pi_{\frac{1}{2}}$ subsystem is also apparent, but is rather weak as a result of the reduced population of the lower level because of collisional deactivation.

The combination of the information forthcoming from the excitation and emission (the $\tilde{A}^2\Sigma^+ \rightarrow \tilde{X}^2\Pi_i$ transition is also seen) experiments allows most of the features in the two spectra to be assigned and results in the determination of the vibrational frequencies of ^{35}ClCN$^+$ and ^{37}ClCN$^+$ in the $\tilde{X}^2\Pi_i$, $\tilde{A}^2\Sigma^+$ and $\tilde{B}^2\Pi_i$ states, as well as the spin–orbit constants. Most recently, the rotational structure of some of the bands has been resolved by using narrower laser bandwidth ($ca.$ 0.04 cm^{-1}) excitation and the analysis of such data for various isotopically labelled derivatives of ClCN$^+$ leads to their rotational constants and r_s geometric structure (Celii $et\ al.$ 1988). Complementary vibronic and rotational analyses have also been accomplished for the isotopes of BrCN$^+$ (Hanratty $et\ al.$ 1988).

The haloacetylene cations, XCCH$^+$, have often featured as prototypes in the development of the spectroscopic techniques for open-shell cations. Following the detection of their $\tilde{A}^2\Pi_i \rightarrow \tilde{X}^2\Pi_i$ emission spectra by electron-impact excitation on an effusive beam (Allan $et\ al.$ 1977), the vibronic analysis of this transition had to await the introduction of the supersonic free jet and the laser excitation approaches before the complexity of the spectra could be disentangled (Maier 1986). In particular, in an analogous way to that illustrated above for ClCN$^+$, the $\tilde{A}^2\Pi_i \leftarrow \tilde{X}^2\Pi_i$ laser excitation spectra of XCCH$^+$, X = Cl, Br, I, recorded with moderate resolution (*ca.* 0.2 cm^{-1}) enabled the origins to be identified and this led, in turn, to vibrational analysis of the band systems. The vibrational frequencies are thus known to ± 1 cm^{-1} for most of the modes for these ions in their $\tilde{X}^2\Pi_i$ and $\tilde{A}^2\Pi_i$ states.

Resolution of the rotational structure in the laser excitation spectra of the haloacetylene cations XCCH$^+$, X = Cl (King *et al.* 1985*b*), Br (King *et al.* 1984*b*), and I (Maier & Ochsner 1985) followed, and the rotational and related spectroscopic constants have been evaluated. Because in all three haloacetylene cations only the $\tilde{A}^2\Pi_\frac{3}{2} \leftarrow \tilde{X}^2\Pi_\frac{3}{2}$ spin-orbit subsystem could be studied, the analyses yield the B_{eff} constants. These are collected in table 2 together with the B_0 values for the diacetylene cations obtained also by this approach from a simultaneous fit for $\Omega = \frac{3}{2}$ and $\frac{1}{2}$ components (Kuhn *et al.* 1986).

In figure 6 the rotationally resolved laser excitation spectrum of the 3^1_0 band of the $\tilde{A}^2\Pi_\frac{3}{2} \leftarrow \tilde{X}^2\Pi_\frac{3}{2}$ transition of iodoacetylene cation is reproduced (Maier & Ochsner 1985). The R- and P-branches are designated; the Q-branch is too weak to be detected at the ambient temperature 100–150 K employed in the measurements. The spectroscopic constants were determined by a least-squares fit between the line positions and those calculated from the differences of the eigenvalues of the upper- and lower-state hamiltonians. The effective closed-form expression used for the rotational energy levels,

$$F_\nu(J) = B_{\text{eff},\nu}([J+\tfrac{1}{2}]^2 - 1) - D_{\text{eff},\nu}([J+\tfrac{1}{2}]^2 - 2)^2,$$

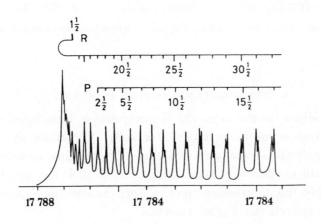

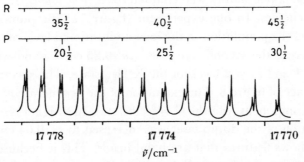

FIGURE 6. Rotationally resolved laser excitation spectrum (0.04 cm^{-1} FWHM) of the $\tilde{A}^2\Pi_\frac{3}{2} \leftarrow \tilde{X}^2\Pi_\frac{3}{2}$ 3^1_0 transition of the deuteroiodoacetylene cation.

[141]

is obtained by second-order perturbation theory from the published matrix elements (Zare *et al.* 1973). The determined rotational constant B_{eff} by this treatment is given in table 2 for the zeroth level of the $\tilde{X}^2\Pi$ and $\tilde{A}^2\Pi$ states of ICCH$^+$ and ICCD$^+$.

For the chloroacetylene and bromoacetylene cations, the presence of the naturally occurring halogen isotopes complicates the spectra on the one hand, but in return yields the partial r_s structure from the knowledge of the respective rotational constants.

TABLE 2. ROTATIONAL CONSTANTS (RECIPROCAL CENTIMETRES) FOR THE HALOACETYLENE AND DIACETYLENE CATIONS INFERRED FROM THE STRUCTURE IN THEIR $\tilde{A}^2\Pi_i \leftarrow \tilde{X}^2\Pi_i$ LASER EXCITATION SPECTRA[a]

cation	$B_{0,\,eff}$	
	$\tilde{X}^2\Pi_{\frac{3}{2}}$	$\tilde{A}^2\Pi_{\frac{3}{2}}$
^{35}Cl—C≡C—H$^+$	0.194 647 (49)	0.170 881 (48)
^{37}Cl—C≡C—H$^+$	0.191 063 (60)	0.167 680 (60)
^{35}Cl—C≡C—D$^+$	0.176 968 (38)	0.156 338 (37)
^{37}Cl—C≡C—D$^+$	0.173 665 (46)	0.153 342 (45)
^{79}Br—C≡C—H$^+$	0.137 794 (37)	0.121 351 (36)
^{81}Br—C≡C—H$^+$	0.137 327 (38)	0.121 032 (38)
^{79}Br—C≡C—D$^+$	0.125 767 (29)	0.111 312 (29)
^{81}Br—C≡C—D$^+$	0.125 152 (33)	0.110 772 (33)
I—C≡C—H$^+$	0.109 59 (7)	0.096 69 (7)
I—C≡C—D$^+$	0.100 44 (8)	0.088 99 (8)
	B_0	
	$\tilde{X}^2\Pi$	$\tilde{A}^2\Pi$
H—(C≡C)$_2$—H$^+$	0.146 90 (4)	0.140 09 (4)
D—(C≡C)$_2$—D$^+$	0.127 62 (5)	0.122 03 (5)

[a] The references to the studies are given in the text. The values in brackets correspond to one standard deviation.

4. STIMULATED EMISSION PUMPING

The most recent addition to the armoury of spectroscopic techniques available to probe cations is stimulated emission pumping (Kittrell *et al.* 1981). Previously, this approach has been used in the study of highly vibrationally excited levels of closed-shell molecules in their ground electronic state in a static environment (Hamilton *et al.* 1986). It has now been demonstrated that one can also apply this to transient species, such as ions, generated in much smaller concentrations in flow systems (Celii *et al.* 1986 b).

The ion used to test the feasibility of this approach was the diacetylene cation, for which the $\tilde{A}^2\Pi_u \leftrightarrow \tilde{X}^2\Pi_g$ transition is reasonably well characterized (*vide supra*). In figure 7 are depicted the optical pumping schemes. In one experiment (figure 7 a) the pump laser (*ca.* 0.08 cm^{-1} band-width) transferred the population to the lowest vibrational level of the $\Omega = \frac{3}{2}$ component of the $\tilde{A}^2\Pi_u$ state whereas the second dye laser (*ca.* 0.25 cm^{-1} band-width) stimulated the transitions to the $\nu_3'' = 1$ and $\nu_7'' = 2$ levels of the $\tilde{X}^2\Pi_{\frac{3}{2},\,g}$ state. The spectrum obtained can be seen in the left-hand trace of figure 8. The excitation of the degenerate ν_7'' mode in two quanta results in two vibronic components (for the $\Omega = \frac{3}{2}$ substate). In a higher resolution measurement (figure 7 b), the resolution of the dump laser was increased to *ca.* 0.04 cm^{-1} and the recorded spectrum (figure 8 a) shows features that are band heads. This is because the excitation laser frequency was tuned to correspond to the R-branch heads and consequently *ca.* 6–10 rotational

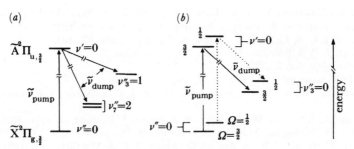

FIGURE 7. Transitions between the ground and excited electronic state of diacetylene cation involved in the stimulated emission pumping experiments.

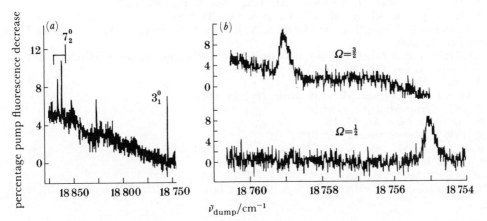

FIGURE 8. Stimulated emission pumping transitions of diacetylene cation (see figure 7) observed (a) at low resolution (0.25 cm^{-1}), and (b) at higher resolution (0.04 cm^{-1}).

levels (around $J = 25.5$) are populated in either the $\Omega = \frac{1}{2}$ or $\Omega = \frac{3}{2}$ component of the $\tilde{A}^2\Pi_u$ state. As a matter of fact, the assignment of the features in the region of the 3_1^0, 7_2^0 transitions was not clear in the emission spectrum (Callomon 1956); the double-resonance nature of the stimulated emission pumping experiment proves which peaks belong to the 3_1^0 (figure 8b) and which belong to the 7_2^0 transition.

A method now exists for probing the ground state of the cations in selected vibrational levels at high resolution to complement such studies in the excited electronic state accessible by the established laser excitation approach.

The research studies described here were realized because of the efforts of the various co-workers whose names are given in the references to the respective studies. The project is financed by the Schweizerischer Nationalfonds zur Förderung der wissenschaftlichen Forschung (no. 2.429-0.84).

REFERENCES

Allan, M. & Maier, J. P. 1976 Chem. Phys. Lett. **41**, 231.
Allan, M., Kloster-Jensen, E. & Maier, J. P. 1977 J. chem. Soc. Faraday Trans. II **73**, 1406.
Callomon, J. H. 1956 Can. J. Phys. **34**, 1046.
Celii, F. G., Fulara, J., Maier, J. P. & Rösslein, M. 1986a Chem. Phys. Lett. **131**, 325.
Celii, F. G., Maier, J. P. & Ochsner, M. 1986b J. chem. Phys. **85**, 6230.

Celii, F. G., Rösslein, M., Hanratty, M. A. & Maier, J. P. 1988 *Molec. Phys.* (In the press.)

Fulara, J., Klapstein, D., Kuhn, R. & Maier, J. P. 1985 *J. phys. Chem.* **89**, 4213.

Hamilton, C. E., Kinsey, J. L. & Field, R. W. 1986 *A. Rev. phys. Chem.* **37**, 493.

Hanratty, M. A., Maier, J. P. & Rösslein, M. 1988 (In preparation.)

Herzberg, G. 1971 *Rev. chem. Soc.* **25**, 201.

Herzberg, G. 1985 *Proc. Indian natn. Sci. Acad.* A **51**, 495.

King, M. A., Klapstein, D., Kroto, H. W., Kuhn, R., Maier, J. P. & Nixon, J. F. 1984*a J. chem. Phys.* **80**, 2332.

King, M. A., Klapstein, D., Kroto, H. W., Maier, J. P. & Nixon, J. F. 1982 *J. molec. Struct.* **80**, 23.

King, M. A., Klapstein, D., Kuhn, R., Maier, J. P. & Kroto, H. W. 1985*a Molec. Phys.* **56**, 871.

King, M. A., Kroto, H. W., Nixon, J. F., Klapstein, D., Maier, J. P. & Marthaler, O. 1981 *Chem. Phys. Lett.* **82**, 543.

King, M. A., Kuhn, R. & Maier, J. P. 1986 *J. Phys. Chem.* **90**, 6460.

King, M. A., Kuhn, R. & Maier, J. P. 1987 *Molec. Phys.* **60**, 867.

King, M. A., Maier, J. P., Misev, L. & Ochsner, M. 1984*b Can. J. Phys.* **62**, 1437.

King, M. A., Maier, J. P. & Ochsner, M. 1985*b J. chem. Phys.* **83**, 3181.

Kittrell, C., Abramson, E., Kinsey, J. L., McDonald, S. A., Reisner, D. E., Field, R. W. & Katayama, D. H. 1981 *J. chem. Phys.* **75**, 2056.

Klapstein, D., Maier, J. P. & Misev, L. 1983 In *Molecular ions: spectroscopy, structure and chemistry* (ed. T. A. Miller & V. E. Bondybey), pp. 175–200. New York: North Holland.

Kroto, H. W. 1982 *Chem. Soc. Rev.* **11**, 435.

Kuhn, R., Maier, J. P. & Ochsner, M. 1986 *Molec. Phys.* **59**, 441.

Maier, J. P. 1980 *Chimia* **34**, 219.

Maier, J. P. 1982 *Acct. Chem. Res.* **15**, 18.

Maier, J. P. 1986 *J. electron. Spectrosc.* **40**, 203.

Maier, J. P. & Ochsner, M. 1985 *J. chem. Soc. Faraday Trans.* II **81**, 1587.

McDonald, J. K. & Innes, K. K. 1969 *J. molec. Spectrosc.* **29**, 251.

Miller, T. A. & Bondybey, V. E. 1982 *Appl. spectrosc. Rev.* **18**, 105.

Zare, R. N., Schmeltekopf, A. L., Harrop, W. I. & Albritton, D. L. 1973 *J. molec. Specrosc.* **46**, 37.

Discussion

E. HIROTA (*Institute for Molecular Science, Okazaki, Japan*). I have two questions on halogenated acetylene cations. First, does Professor Maier get any information on the Renner–Teller effect? Second, why did Professor Maier not study the fluorinated species? We have investigated the HCCO radical which is an isoelectronic molecule of $HC{\equiv}CF^+$, by microwave spectroscopy. We found the radical to be nonlinear and to have a low-lying electronic state.

J. P. MAIER. We have not studied the Renner–Teller effects in the optical spectra of the haloacetylene cations. The $\tilde{A}^2\Pi \leftrightarrow \tilde{X}^2\Pi$ electronic transitions of $XCCH^+$ and $XCCX^+$, X = Cl, Br, I, have been observed both in emission and by laser excitation. We have analysed the main vibrational features of the spectra, but have not looked into vibronic intractions in any detail; the emission spectra in particular are complex. We have, however, studied such interactions in the $^2\Pi$ ground state of the triatomic cations HBS^+, DBS^+, HCP^+ and DCP^+, as outlined in my paper.

As far as the fluorinated species are concerned, the electronic spectra of $FCCH^+$ and $FCCF^+$ have not been investigated. On the basis of the photoelectron spectra, one may observe the $\tilde{A}^2\Pi - \tilde{X}^2\Pi$ transitions around 200 nm. In the case of $FCCCN^+$, fluorescence from the lowest excited electronic states could not be detected, whereas the $\tilde{A}^2\Pi_g \rightarrow \tilde{X}^2\Pi_u$ emission spectrum of $FCCCCF^+$ has been observed and vibrationally analysed.

G. DUXBURY (*Department of Physics and Applied Physics, University of Strathclyde, Glasgow, U.K.*). I wish to draw attention to the use of methods for calculating the Renner–Teller coupling in triatomic molecules (Jungen & Merer 1980; Duxbury & Dixon 1981) for modelling the isotopic dependence of rotation constants and vibronic parameters. Recently, Lew & Groleau (1987) have completed their analysis of the emission spectrum of D_2O^+. In table D1, the $D_2O^+(\tilde{X}, \,^2B_1)$ constants, calculated by using the parameters derived by Jungen et al. (1980) from a fit to the H_2O^+ spectrum are compared with those derived from the analysis of the D_2O^+ spectrum. It can be seen that the degree of agreement is very good. From the calculated values of $A^{SO}_{v,k}$ plotted in figure D1 it can be seen that the sawtooth pattern of the spin–orbit

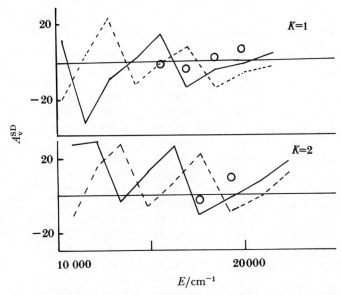

FIGURE D1. Spin–orbit splitting for vibronic levels of the excited state of D_2O^+ with $K = 1$–2. Broken line, calculated as in (b), table D1; solid line, calculated as in (c), table D1.

TABLE D1. ROTATIONAL CONSTANTS (RECIPROCAL CENTIMETRES) OF THE $\tilde{X}$, $^2B_1$
STATE OF D_2O^+

	$v_2'' = 0$	$v_2'' = 1$	$v_2'' = 2$	$v_2'' = 3$	
A	16.0325	17.6955		22.496	(a)
	15.86	17.48	19.71	22.63	(b)
	15.84	17.68	20.17	23.52	(c)
B	6.2399	6.2624		6.310	(a)
	6.212	6.220	6.220	6.218	(b)
	6.203	6.193	6.173	6.153	(c)
C	4.4066	4.3538		4.224	(a)
	4.373	4.317	4.263	4.211	(b)
	4.373	4.318	4.265	4.214	(c)
ϵ_{aa}	−0.586	−0.782		−1.487	(a)
	−0.65	−0.80	−1.01	−1.34	(b)
	−0.68	−0.84	−1.10	−1.50	(c)
G_0	0.0	1044.27		3058.66	(a)
	0.0	1036.81	2044.89	3025.45	(b)
	0.0	1044.13	2054.40	3032.20	(c)

(a) Observed (Lew & Groleau 1987).
(b) Calculated (Jungen et al. 1980).
(c) Calculated as (b) except $H'' = 8648$ cm^{-1} and $f_m'' = 30\,553.90$ cm^{-1} rad^{-2}.

coupling constants is very sensitive to the upper-state – lower-state splitting pattern. The second calculation with an adjusted ground-state potential barrier can be seen to give good qualitative agreement with the data. The vibronic origins listed in table D2 are also in good agreement. To achieve better quantitative agreement it would be necessary to carry out a simultaneous fit to the data on both isotopic species.

TABLE D2. VIBRONIC CONSTANTS (RECIPROCAL CENTIMETRES) FOR THE LEVELS OF
THE $\tilde{A}$, $^2A_1$ STATE OF D_2O^+

	V				
K	bent	linear	T_V	B_V	
0	5	11	15882.32	4.478	(a)
			845.09	4.426	(b)
	6	13	17331.93	4.478	(a)
			341.76	4.453	(b)
	7	15	18807.28	4.468	(a)
			833.38	4.480	(b)
	8	17	20301.03	4.638	(a)
			344.93	4.507	(b)
1	4	10	15156.86	4.88	(a)
			92.60	4.42	(b)
	5	12	16596.14	4.57	(a)
			577.91	4.55	(b)
	6	14	18057.18	4.55	(a)
			45.70	4.48	(b)
	7	16	19540.39	4.45	(a)
			547.75	4.50	(b)
2	5	13	17300.22	4.466	(a)
			286.29	4.45	(b)
	6	15	18706.91	4.55	(a)
			761.32	4.48	(b)
3	4	12	16541.38	4.72	(a)
			570.10	4.47	(b)
	5	14	17989.62	4.57	(a)
			984.03	4.49	(b)

(a) Observed.
(b) Calculated as in (c) table D1.

The isoelectronic molecules CH_2 and NH_2^+ possess a $^3B_1(^3\Sigma)$ ground state with a low-lying singlet state, $^1A_1$, which forms the lower component of a pair that correlate with $^1\Delta$ of linear $CH_2(NH_2^+)$. Recently experiments have been done on CH_2 to measure the magnetic activity in the singlet system Petek et al. (1987). This magnetic activity can result from two causes, residual orbital angular momentum associated with the $^1\Delta$ state, and singlet–triplet mixing. In this discussion I am concentrating on the first of these, angular momentum associated with the linear molecule delta state.

For the dihydrides NH_2 and H_2O^+, a comprehensive study by Jungen & Merer (1980) has shown that the effects of orbital angular momentum can be seen in the erratic variation for small to large, regular or inverted, of the spin–orbit coupling constants of the vibronic levels. This behaviour is associated with the combination of large-amplitude bending vibrational motion with Renner–Teller coupling between the two states, $^2B_1$ and $^2A_1$, which correlate with the $^2\Pi$ state of linear NH_2. In the NH_2 and H_2O^+ this spin–orbit coupling constant $A_{v,k}^{so} = A^{so}\langle\hat{L}_z\rangle$. For a Π state, the maximum value of $\langle\hat{L}_z\rangle = 1$. For CH_2 the maximum value of $\langle\hat{L}\rangle = 2$. However, as the linear molecule limit is $^1\Delta$, there is no spin–orbit interaction constant. The g value of the ro-vibronic levels will be related to the expected value of the orbital angular momentum, i.e.

$$g_{\text{eff}}(K, J) = -K/J\langle\hat{L}_z\rangle = g_r^a K^2/J,$$

where
$$g_r^a = \langle L_z\rangle/K.$$

Christian Jungen and I have recently calculated the variation of $\langle L_z\rangle$ in the singlet levels of CH_2. The result for $K = 1$ is shown in figure D2. It can be seen that the levels recently studied by the Berkeley group, labelled in the figure by M (Petek $et\ al.$ 1987), are those that should show only small effects, and hence most of the magnetic activity observed is because of singlet–triplet mixing.

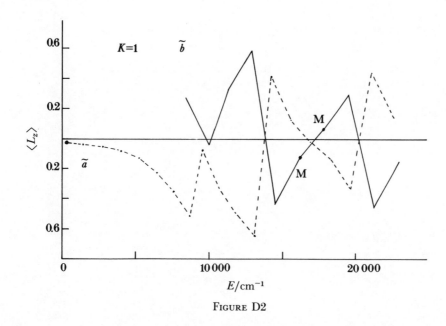

FIGURE D2

Singlet–triplet mixing will, however, depend upon $\langle L_z\rangle$, and our next task is to evaluate explicitly the effects of $\langle L_z\rangle$ on the rate of intersystem crossing. Although NH_2^+ has not been studied in the same detail (Dunlavey $et\ al.$ 1980; Gibson $et\ al.$ 1985), once high-resolution spectra become available similar effects of orbital angular momentum should be shown.

References

Dunlavey, S. J., Dyke, J. M., Jonathan, N. & Morris, A. 1980 *Molec. Phys.* **39**, 1121–1135.
Duxbury, G. & Dixon, R. N. 1981 *Molec. Phys.* **43**, 255–274.
Gibson, S. T., Greene, J. P. & Berkowitz, J. 1985 *J. chem. Phys.* **83**, 4319–4328.
Jungen, Ch., Hallin, K. E. J. & Merer, A. J. 1980 *Molec. Phys.* **40**, 25–94.
Jungen, Ch. & Merer, A. J. 1980 *Molec. Phys.* **40**, 1–24.
Lew, H. 1976 *Can. J. Phys.* **54**, 2028–2049.
Lew, H. & Groleau, R. 1987 *Can. J. Phys.* (Submitted.)
Petek, H., Nesbitt, D. J., Darwin, D. C. & Moore, C. B. 1987 *J. chem. Phys.* **86**, 1172–1188.

Figure 6a

Phil. Trans. R. Soc. Lond. A **324**, 223–232 (1988)

Printed in Great Britain

Fluorescence excitation spectroscopy of ionic clusters containing the $C_6F_6^+$ chromophore

By C.-Y. Kung[1], T. A. Miller[1] and R. A. Kennedy[2]

[1] *The Ohio State University Laser Spectroscopy Facility, Department of Chemistry,*
The Ohio State University, 120 West 18th Avenue, Columbus, Ohio 43210, U.S.A.

[2] *Department of Chemistry, Birmingham University, P.O. Box 363, Birmingham B15 2TT, U.K.*

Fluorescence excitation spectra of the ion–molecule complexes $C_6F_6^+ \cdot X_n$ (X = He, Ne, Ar or N_2) are reported. It is shown that such data can be used to test model potentials for these complexes. For $n = 1$ and X = He and Ne, a simple long-range model of the interaction potential is nearly adequate, but for X = Ar it appears that a substantial charge-transfer component to the interaction is required.

Introduction

The spectroscopy of ion–molecule clusters potentially offers a very direct probe of the interactions between an ion and one or more neutral species. An understanding of these forces is crucial to the development of models for such fundamental processes as nucleation about ions (e.g. the cloud chamber) and the solvation of ions. Ionic clusters are also known to be involved in the chemistry of the upper atmosphere (Smith & Adams 1980), and play a role in some discharge processes (Kaufman *et al.* 1980).

The success of the spectroscopic approach to the investigation of intermolecular forces is amply demonstrated by the many studies of neutral clusters (Levy 1981; Beswick & Jortner 1981). Although ion–molecule complexes have attracted considerable experimental attention (Märk & Castleman 1985), detailed spectroscopic information concerning them is extremely sparse. Schwarz (1977; 1980) has published low-resolution infrared (IR) spectra of charged H_2O and NH_3 clusters, and Okumura *et al.* (1985, 1986) have observed vibrational predissociation spectra of H_5^+, H_7^+ and H_9^+ and of $H_7O_3^+ \cdot H_2$ and $H_9O_4^+ \cdot H_2$. DiMauro *et al.* (1984) have reported the red shift of the origin of the $\tilde{B}$–$\tilde{X}$ band of $C_6F_6^+$ on complexation by one or more He, Ne or Ar atoms. This paper describes more recent work on the $C_6F_6^+ \cdot X_n$ system in which vibrational structure associated with the motion of the neutral relative to the ion has been observed. These experimental observations are being supported by the development of theoretical methods for the calculation of the vibrational levels of these complexes from model potentials. It will be shown how the combination of the experimental observations and the theoretical calculations is leading to the spectroscopic determination of potentials for these complexes. A preliminary report of work on $C_6F_6^+ \cdot He$ has been published (Kennedy & Miller 1986).

Experimental methods

The intermolecular bonds in neutral and ionic clusters are relatively weak. To obtain adequate quantities of these species for their spectroscopic characterization, it is necessary to work in an environment in which there are few collisions that may break up the clusters after their formation. For neutral clusters the unique properties of supersonic expansions have proved to be ideal (Levy 1981). Precursor molecules of interest are seeded into an inert carrier gas and the mixture expanded through a small orifice. The clusters are formed in the high-pressure region of the expansion close to the nozzle in three-body collisions. Further downstream in the expansion there are very few collisions, and the collision energies are extremely low, being characterized by translational temperatures of only a few kelvin. Once formed, the clusters will not be destroyed until they encounter the background gas in the chamber at a shock front, or the walls of the chamber.

The formation of ion–molecule clusters by this approach is complicated by the fact that molecular ions are highly reactive species. Ions seeded into the carrier gas before the expansion will be destroyed both by ion–molecule reactions and by neutralization on the walls of the metal nozzle before they pass through the expansion orifice. The ions must instead be formed within the supersonic expansion, and neutral species attached in subsequent collisions.

$C_6F_6^+$ was chosen as the ion for these experiments for a number of reasons.

1. The ion is cleanly formed from the neutral by nonresonant photoionization with two photons from an excimer laser operating on the ArF (193 nm, 2 $hv = 104\,000$ cm$^{-1} = 12.8$ eV) transition. The ionization potential of $C_6F_6^+$ is about 9.9 eV.

2. C_6F_6 is easily entrained in the expansion by passing the carrier gas, helium of HP grade ($+$Ne, Ar, N_2 ..., if desired) over the liquid held in a temperature-controlled bath at or below room temperature.

3. C_6F_6 is an excellent chromophore for the observation of fluorescence excitation spectra. The $\tilde{B}$–$\tilde{X}$ system of the bare ion is very strong, has a fluorescence quantum yield of unity, and is easily accessible in the blue region of the spectrum ($\tilde{\nu}_0 = 21\,616$ cm^{-1}).

Details of the experimental method have been given elsewhere (Di Mauro et al. 1984). The formation of ionic clusters depends critically on the point at which photoionization occurs. From the results of experiments in which the number of ionic clusters present in the expansion was monitored as a function of the distance from the tip of the nozzle to the excimer laser crossing point, it appears that a number of steps are involved in the formation of an ion cluster.

1. Formation of a neutral cluster, which occurs very close to the nozzle (within 10 nozzle diameters).

2. Photolysis of the neutral cluster to leave a bare ion. Fragmentation of the neutral cluster during photolysis disposes of the excess energy deposited in the organic molecule, thereby lessening fragmentation of the parent molecule during or after photoionization.

3. Attachment of one or more neutral species to the bare ion in three-body collisions. Note that direct photoionization of a neutral cluster to form an ionic cluster does not appear to be a significant process. The optimum nozzle–excimer distance was found to be about 10 nozzle diameters (ca. 1.5 mm). Moving the nozzle closer to the excimer leads to the appearance of a variety of photolysis fragmentation products including C_2, whereas pulling the nozzle away does not allow the bare ions to undergo three body collisions to form clusters. The distance of 10 nozzle diameters applies for a helium reservoir pressure of ca. 15 atm†. This was found to be the optimum pressure for forming small ionic clusters such as $C_6F_6^+ \cdot$ He. Higher pressures cause more collisions to occur after the bare ion is formed, leading to the formation of larger clusters (Di Mauro et al. 1984).

The best approach to generating $C_6F_6^+ \cdot X$ (e.g. X = Ne, Ar, N_2) is first to produce good

† 1 atm $\approx 10^5$ Pa.

spectra of $C_6F_6^+\cdot He$, and then add a small amount (at most a few percent, by mass) of the species, X, to be attached to $C_6F_6^+$ to the carrier gas. The formation of $C_6F_6^+\cdot X$ may well proceed by displacement of the He atom

$$C_6F_6^+\cdot He + X \rightarrow C_6F_6^+\cdot X + He.$$

Addition of larger quantities of X (especially Ar and N_2) leads to the formation of larger clusters $C_6F_6^+\cdot X_n$, and also quenches the generation of $C_6F_6^+$ by excimer laser photoionization.

SPECTROSCOPY AND THEORY OF $C_6F_6^+\cdot He$

The original spectroscopic investigations of the ion–atom complex $C_6F_6^+\cdot He$ were restricted to the determination of the red shift (ca. 38 cm^{-1}) of the origin of the $\tilde{B} \leftarrow \tilde{X}$ system on attaching a helium atom (Di Mauro et al. 1984). The data on this ion have now been greatly extended through the observation of a vibrational progression in the modes corresponding to the motion of the He atom with respect to the chromophore ion (Kennedy & Miller 1986). The analysis of this progression is leading to the determination of a model potential for this ionic cluster. This experiment is believed to be the first to so-directly probe the potential surface of an ionic cluster.

The experimental observations of Kennedy & Miller provide the location of four excited vibrational levels of $C_6F_6^+(\tilde{B})\cdot He$. This information can be used to test and refine model potentials for the complex. Jortner et al. (1983) have proposed a potential for systems of this type, consisting of a sum of pairwise Lennard–Jones 12–6 interactions for the helium atom with the carbon and fluorine atoms

$$V_{LJ}(\boldsymbol{r}_{He}) = \sum_i A_i \left\{ \frac{-1}{r_{iHe}^6} + \frac{\frac{1}{2}(r_{iHe}^0)^6}{r_{iHe}^{12}} \right\}, \tag{1}$$

where

$$r_{iHe} = |\boldsymbol{r}_i - \boldsymbol{r}_{He}|, \tag{2}$$

and a charge-induced dipole term describing the interaction of the distribution of the net positive charge on the $C_6F_6^+$ ion with the polarizability of the helium atom

$$V_{cid}(\boldsymbol{r}_{He}) = -\tfrac{1}{2}\alpha_{He}e^2 |F(\boldsymbol{r}_{He})|^2, \tag{3}$$

where

$$F_\xi = \sum_i \frac{(e_i/e)}{r_{iHe}^3}(\xi_i - \xi_{He}); \quad \xi = x, y, z. \tag{4}$$

The total potential

$$V_{tot}(\boldsymbol{r}_{He}) = V_{LJ}(\boldsymbol{r}_{He}) + V_{cid}(\boldsymbol{r}_{He}) \tag{5}$$

$$= V(x, y, z) \tag{6}$$

is a function of three spatial variables that were chosen as z, the height of the helium atom above the ring, and the orthogonal displacements (x, y) of the atom from the C_6 axis through the centre of the ring. The x-axis is chosen to lie $15°$ from the direction of a C–F bond, so that the potential cuts $V(x, 0, z)$ and $V(0, y, z)$ are identical.

If the $C_6F_6^+$ ring is assumed to be infinitely massive compared to the helium atom, the hamiltonian describing the motion of the atom relative to the fixed ring may be written as

$$H = \frac{-\hbar^2}{2m_{He}} \left(\frac{d^2}{dx^2} + \frac{d^2}{dy^2} + \frac{d^2}{dz^2} \right) + V(x, y, z). \tag{7}$$

[151]

The assumption is also made that the $C_6F_6^+$ chromophore may be treated as a rigid body. This assumption is supported by the observation that the cluster spectra associated with different vibrational levels of $C_6F_6^+(\tilde{B})$ are identical.

Plots of the total potential, $V(x, y, z)$, with reasonable values for the various parameters show two interesting features (see figure 1). First, the minimum is located above the centre of the

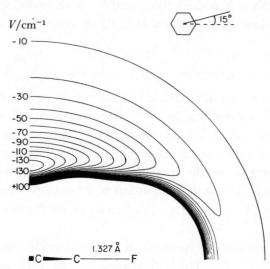

FIGURE 1. Contour plot of the $V(x, 0, z)$ cut of the potential for $C_6F_6^+ \cdot He$ described in the text. The minimum is located 2.43 Å above the plane of the $C_6F_6^+$ ring. Contours are plotted every 10 cm^{-1}.

ring, and secondly, the minimum energy path for moving the helium atom away from the C_6 axis keeps the height of the atom above the ring approximately fixed at the equilibrium value for quite substantial displacements. The hamiltonian (7) may usefully be expressed as the sum

$$H = H_s(z) + H_b(x) + H_b(y) + \Delta V(x, y, z), \tag{8}$$

where the 'stretching' hamiltonian is

$$H_s(z) = \frac{\hbar^2}{2m_{He}} \frac{d^2}{dz^2} + V_s(z), \; V_s(z) = V(0, 0, z) \tag{9}$$

and the 'bending' hamiltonian is

$$H_b(x) = \frac{\hbar^2}{2m_{He}} \frac{d^2}{dx^2} + V_b(x), \; V_b(x) = V(x, 0, z_m), \tag{10}$$

(z_m is such that $(dV_s/dz)_{z=z_m} = 0$, i.e. z_m is the height of the atom above the ring at the minimum of the stretching potential), $H_b(y)$ is similar to $H_b(x)$, and

$$\Delta V(x, y, z) = V(x, y, z) - V_s(z) - V_b(d), \tag{11}$$

where $|d| = (x^2 + y^2)^{\frac{1}{2}}$. The determination of the vibrational energy levels for the model potential $V(x, y, z)$ now breaks down into three stages.

(a) Calculation of the zeroth-order stretching and bending vibrational energy levels and wavefunctions by numerically solving the separated Schrödinger equations for the vibrational coordinates z and x (or y)

$$\left[\frac{-\hbar^2}{2m_{He}}\frac{d^2}{dz^2} + V_s(z)\right]X_s(z) = E_s X_s(z) \tag{12}$$

and

$$\left[\frac{-\hbar^2}{2m_{He}}\frac{d^2}{dx^2} + V_b(x)\right]X_b(x) = E_b X_b(x). \tag{13}$$

These equations were solved by using the Numerov–Cooley algorithm (Cooley 1961).

(b) Evaluation of the first-order corrections, E', to the combined stretching and bending energy levels from the difference potential perturbation, ΔV,

$$E' = \langle sb_1 b_2 | \Delta V | sb_1 b_2 \rangle, \tag{14}$$

where

$$(sb_1 b_2) = X_s(z) X_{b_1}(x) X_{b_2}(y). \tag{15}$$

The required matrix elements were evaluated by three-dimensional numerical quadrature with Gauss–Legendre weights and abscissae; up to $192 \times 192 \times 192$ points were employed to ensure convergence.

(c) Evaluation of the off-diagonal matrix elements of ΔV:

$$\Delta V^{mn} = \langle s^m b_1^m b_2^m | \Delta V | s^n b_1^n b_2^n \rangle \tag{16}$$

followed by the construction and diagonalization of the numerically generated hamiltonian matrix. The matrix elements were again computed by three-dimensional Gauss–Legendre quadrature. Symmetry considerations show that off-diagonal matrix elements with Δb_1 and/or Δb_2 odd must be zero. For levels that are doubly degenerate through first order, $(sb_1 b_2)$, $(sb_2 b_1)$, $b_1 \neq b_2$, symmetric and antisymmetric combinations can be formed

$$\sqrt{\tfrac{1}{2}}[(sb_1 b_2) + (sb_2 b_1)] \text{ and } \sqrt{\tfrac{1}{2}}[(sb_1 b_2) - (sb_2 b_1)].$$

The transition observed in the fluorescence excitation spectrum from the vibrationless level of the ground state must be to totally symmetric vibrational levels of the excited state. These levels are constructed from the singly degenerate basis functions $(s00)$ and $(s2b_1 2b_1)$, together with the symmetric combination of the stretch-bend functions with both b_1 and b_2 even, i.e. $\sqrt{\tfrac{1}{2}}[(s2b_1 2b_2) + (s2b_2 2b_1)]$. The hamiltonian matrix for these levels is set up, including basis functions up to a fixed energy above (000), and diagonalized.

The energies obtained at the different stages in the calculation are reported in table 1, together with the experimental results for comparison. The parameters used in generating the model potential were obtained from various sources. The geometry of $C_6F_6^+(\tilde{B})$ was assumed to be the same as for the ground state of neutral $C_6F_6^+$, $r_{CC} = 1.394$ Å†, determined by electron

TABLE 1. RESULTS OF THEORETICAL CALCULATIONS ON THE VIBRATIONAL ENERGY LEVELS OF $C_6F_6^+ \cdot He$

level $(sb_1 b_2)$	zeroth-order energy/cm⁻¹	first-order energy/cm⁻¹	energy from diagonalization/cm⁻¹	experimental/cm⁻¹
(000)	0.0	0.0	0.0	0.0
(100)	48.5	45.3	47.7	41.5
(020) + (002)	43.6	44.0	33.8	48.4
(040) + (004)	67.0	66.5	57.4	66.4
(200)	77.9	70.8	72.3	75.3

† 1 Å $= 10^{-10}$ m $= 10^{-1}$ nm.

diffraction (Almenningen *et al.* 1964). The polarizability of the He atom is $\alpha_{He} = 0.204956 \times 10^{-24}$ cm^3 (Miller & Bederson 1977). The Lennard–Jones parameters for C–He and F–He were extrapolated from those used for previous model calculations on aromatic – rare-gas systems (Ondrechen *et al.* 1981), $A_C = 4000$ cm^{-1}, $r^0_{CHe} = 3.3$ Å, $A_F = 3800$ cm^{-1}, $r^0_{FHe} = 3.2$ Å. The net positive charge on the $C_6F_6^+$ ion was distributed according to the scheme proposed by Jortner *et al.* (1983); $e_C/e = \frac{1}{6}$, $e_F/e = 0$. Basis functions up to 120 cm^{-1} above the (000) level were included in the construction of the full hamiltonian matrix.

The comparison of the calculations with the experimental results is quite appealing. The most obvious discrepancy is that the bending level (002) + (020) is predicted to lie below the stretching level (100), whereas the experimental results suggested the reverse order. Attempts at reversing the order predicted by theory through variations in the four Lennard–Jones parameters have so far proved to be unsuccessful. A relatively small shift of the zeroth-order levels is required; once (100) is below (002) + (020) the substantial Fermi resonance between these levels will push (002) + (020) up still further. The calculations did reveal the anticipated Fermi resonance between the pairs of levels, (100) with (002) + (020) and (200) with (004) + (040), through which transitions to the bending levels borrow intensity from the allowed stretching progression. Inspection of the eigenvector coefficients obtained by diagonalization of the hamiltonian matrix showed that there was also substantial mixing of (002) + (020) with (102) + (120) and of (004) + (040) with (104) + (140). It is believed that these mixings reflect the fact that there is a tendency for the height of the He atom above the ring to decrease as it moves away from the centre (see the contour plot of the potential in figure 1). In the limit, the He atom may orbit the ion, making the method used in these calculations inadequate.

The minimum of the model potential described in the text above is located 2.43 Å above the plane of the ring; the equilibrium-well depth is 136 cm^{-1}. The energies of the vibrational levels that we have observed amount to a very substantial fraction of the well depth, and the potential surface is extensively sampled. The red shift of the cluster origin leads to the conclusion that the well depth for $C_6F_6^+(\tilde{X}) \cdot$ He is about 38 cm^{-1} shallower than for $C_6F_6^+(\tilde{B}) \cdot$ He, indicating a well depth of *ca.* 100 cm^{-1} for $C_6F_6^+(\tilde{X}) \cdot$ He.

Spectroscopy of $C_6F_6^+ \cdot$ Ne, $C_6F_6^+ \cdot$ Ar and $C_6F_6^+ \cdot$ N$_2$

In addition to $C_6F_6^+ \cdot$ He, a number of other ion–atoms and ion–molecule complexes containing the $C_6F_6^+$ chromophore have been investigated.

Figure 2 shows the spectrum of $C_6F_6^+ \cdot$ Ne around the origin of the bare ion, and for comparison the same region for $C_6F_6^+ \cdot$ He. The pattern of lines for $C_6F_6^+ \cdot$ Ne is very similar to that for $C_6F_6^+ \cdot$ He. It is possible that there is a transition of $C_6F_6^+ \cdot$ Ne concealed beneath the origin of bare $C_6F_6^+$. Table 2 tabulates the results for $C_6F_6^+ \cdot$ Ne obtained experimentally, and also gives the results of a preliminary calculation of the vibrational energy levels (parameters $A_C = 11100$ cm^{-1}, $r^0_C = 3.49$ Å, $A_F = 5000$ cm^{-1}, $r^0_F = 3.20$ Å, $\alpha_{Ne} = 0.395 \times 10^{-24}$ cm^3). This potential gives a well depth of *ca.* 210 cm^{-1}, with the Ne atom 2.66 Å above the ring. The match between experiment and theory is again appealing, suggesting that the model potential constructed from considerations of the purely physical interactions within the complex provides a useful description of the system.

A long progression of sharp transitions due to $C_6F_6^+ \cdot$ Ar can be recorded by adding *ca.* 1 % Ar to the He carrier gas (see figure 3). The preliminary conclusions from the spectra

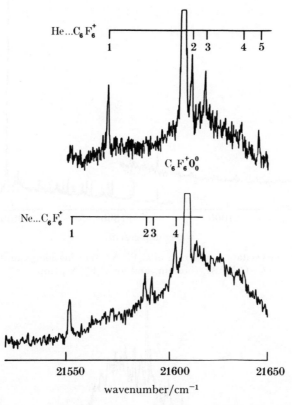

FIGURE 2. Fluorescence excitation spectra of $C_6F_6^+ \cdot He$ and $C_6F_6^+ \cdot Ne$ associated with the origin transition of $C_6F_6^+$.

TABLE 2. $C_6F_6^+ \cdot Ne$ EXPERIMENTAL OBSERVATIONS AND PRELIMINARY RESULTS OF THEORETICAL CALCULATIONS

identification	shift relative to bare-ion origin/cm^{-1}	shift relative to cluster origin/cm^{-1}	theory	vibrational energy level
1	−57.4	0.0	0.0	(000)
2	−20.6	36.8	34.0	(100)
3	−17.4	40.0	26.39	(020)
4	−6.5	50.9	48.79	(040)
			63.42	(200)

of $C_6F_6^+ \cdot Ar$ are that the dramatic increase in the red shift on going from He (37.5 cm^{-1}) or Ne (57.4 cm^{-1}) to Ar (420 cm^{-1}), and the observed increase in the stretching vibrational spacing to *ca.* 66 cm^{-1} indicate that the binding between an Ar atom and $C_6F_6^+$ ($\tilde{B}$) is much stronger than for the lighter rare gases. Model calculations of the type developed for $C_6F_6^+ \cdot He$ support this conclusion, with reasonable values for the Lennard–Jones parameters and the known polarizability of the Ar atom (Miller & Bederson 1977), the stretching frequency is grossly underestimated (calculated 44 cm^{-1}). The increased strength of the attractive interactions between Ar and $C_6F_6^+$ ($\tilde{B}$), compared to the essentially physical binding of the lighter rare gases to this ion, arises from the development of some charge-transfer bonding in the $C_6F_6^+ \cdot Ar$ complex, which is facilitated by the lower ionization potential of Ar relative to He and Ne.

Experiments have also been conducted to try to attach molecules, rather than atoms, to

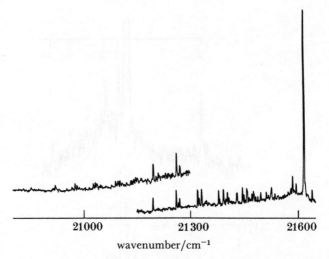

FIGURE 3. Fluorescence excitation spectrum of $C_6F_6^+ \cdot Ar_n$ showing long vibrational progressions for $C_6F_6^+ \cdot Ar$, (bottom) and for $C_6F_6^+ \cdot Ar_2$ (top).

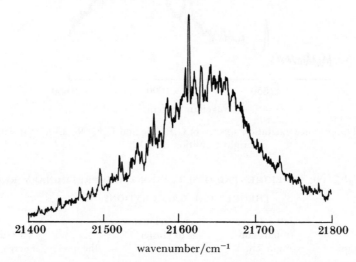

FIGURE 4. Fluorescence excitation spectrum observed close to the origin of $C_6F_6^+$ with a He carrier gas containing *ca.* 2 % N_2 (by volume).

$C_6F_6^+$. By using a He carrier gas containing a small amount of N_2 the spectrum shown in figure 4 was obtained. The series of lines starting 200 cm^{-1} to the red of $C_6F_6^+ \binom{0}{0}$ is tentatively attributed to $C_6F_6^+ \cdot N_2$. The same series of lines can be discerned associated with the 18_0^1 and 17_0^1 transitions of $C_6F_6^+$. As can be seen from the figure, this complex shows a rather more complicated pattern of vibrational structure than seen for the $C_6F_6^+ \cdot X$ (X = He, Ne or Ar) clusters. The additional complexity presumably arises from hindered motions of the N_2 within the cluster. The possibility that some of the features are caused by higher clusters $C_6F_6^+ \cdot (N_2)_n$ has not been definitely eliminated. One curious feature of this spectrum, when compared to the $C_6F_6^+$–rare-gas clusters, is that the intensity of the cluster vibrational transitions at first shows a considerable, steady increase as the vibrational energy in the cluster modes is raised. This suggests that there is a considerable change in the equilibrium distance of the N_2 from the $C_6F_6^+$ and/or in the equilibrium geometry of the cluster when the $C_6F_6^+$ unit is electronically excited.

[156]

The generation of a model potential for the $C_6F_6^+ \cdot N_2$ system must take account of the anisotropy of the polarizability of N_2 and its quadrupole moment, in addition to the terms already described for $C_6F_6^+ \cdot He$. A suitable potential form has not yet been established. The investigation of the vibrational dynamics of the N_2 molecule relative to the $C_6F_6^+$ ion are also more involved than for $C_6F_6^+ \cdot He$. Two new degrees of freedom, hindered rotations of the N_2 unit within the complex, have been added to the problem. The N–N vibration is considered to be of too high a frequency to significantly influence the cluster modes.

CONCLUSIONS

The detailed characterization of the forces between molecules is best accomplished through the spectroscopy of clusters held together by these forces. From the vibrational structure in the fluorescence excitation spectra of $C_6F_6^+ \cdot X$ ion–molecule complexes, it is possible to obtain a substantial amount of data relevant to the ion–molecule forces between a $C_6F_6^+$ ion and a neutral X. These data may be used to test model potentials for these systems. For X = He or Ne, it appears that the results may be adequately accounted for by a purely physical description of the interactions between the ion and the atom. For X = Ar, there is compelling evidence for a substantial charge-transfer component to the binding.

R.A.K. gratefully acknowledges the award of a NATO–SERC postdoctoral fellowship, while working at The Ohio State University. This work was supported by the National Science Foundation under grant CHE-8507537.

REFERENCES

Almenningen, A., Bastiansen, O., Seip, R. & Seip, H. M. 1964 *Acta chem. scand.* **18**, 2115.
Beswick, J. A. & Jortner, J. 1981 *Adv. chem. Phys.* **47**, 363.
Cooley, J. W. 1961 *Math. Comput.* **15**, 363.
DiMauro, L. F., Heaven, M. & Miller, T. A. 1984 *Chem. Phys. Lett.* **104**, 526.
Jortner, J., Even, U., Leutwyler, S. & Berkovitch-Yellin, Z. 1983 *J. chem. Phys.* **78**, 309.
Kaufman, Y., Avivi, P., Dothan, F., Keren, H. & Malinowitz, J. 1980 *J. chem. Phys.* **72**, 2606.
Kennedy, R. A. & Miller, T. A. 1986 *J. chem. Phys.* **85**, 2326.
Levy, D. H. 1981 *Adv. chem. Phys.* **47**, 323.
Märk, T. D. & Castleman, A. W. Jr 1985 *Adv. atom. molec. Phys.* **20**, 66.
Miller, T. M. & Bederson, B. 1977 *Adv. atom. molec. Phys.* **13**, 1.
Okumura, M., Yeh, L. I. & Lee, Y. T. 1985 *J. chem. Phys.* **83**, 3705.
Okumura, M., Yeh, L. I., Myers, J. D. & Lee, Y. T. 1986 *J. chem. Phys.* **85**, 2328.
Ondrechen, M. J., Berkovitch-Yellin, Z. & Jortner, J. 1981 *J. Am. chem. Soc.* **103**, 6586.
Schwartz, H. A. 1977 *J. chem. Phys.* **67**, 5525.
Schwartz, H. A. 1980 *J. chem. Phys.* **72**, 284.
Smith, D. & Adams, N. G. 1980 *Top. curr. Chem.* **89**, 1.

Discussion

H.-J. FOTH (*University of Kaiserlauten, F.R.G.*). Professor Miller described the low temperature obtained in the adiabatic expansion of the supersonic molecular beam. What can be said about the temperature of the ions? Is photoionization a soft ionization, or are the ionic species heated?

T. A. MILLER. It is hard to make a statement that is true for all ions. However, generally speaking, the internal temperature of the ion will approach the translational temperature of the expansion. The example of the CO^+ that was discussed, is typical; its LIF spectrum is indicative of a rotational temperature of a few kelvin.

Conservation of angular momentum requires that there be little difference in the J-states of the precursor and its ion (assuming no fragmentation). Often, the potential function of an ion is sufficiently similar to its parent neutral, that Frank–Cordon considerations dictate little or no change in the vibrational quantum numbers. Therefore, assuming a 'cold' precursor molecule, the ion is also likely to be 'cold'.

Phil. Trans. R. Soc. Lond. A **324**, 233–246 (1988)

Printed in Great Britain

Laser photofragment spectroscopy of near-threshold resonances in SiH+

By P. J. Sarre, J. M. Walmsley and C. J. Whitham

Department of Chemistry, University of Nottingham, University Park,
Nottingham NG7 2RD, U.K.

Near-threshold resonances in the photodissociation cross section of SiH+ have been recorded by the detection of photofragment Si+ ions. The photodissociation spectrum is found to be dominated by Feshbach resonances. The first evidence for multichannel resonances in the photodissociation spectrum of a diatomic molecule is presented.

Laser photodissociation spectra between $15\,600\ \mathrm{cm^{-1}}$ and $18\,750\ \mathrm{cm^{-1}}$ were recorded at a resolution of $0.0012\ \mathrm{cm^{-1}}$ by coaxial laser irradiation of a fast ion beam of SiH+. Over 70 transitions were observed, the majority of which involve excited-state levels (resonances) that lie between the $\mathrm{Si^+(^2P_{\frac{3}{2}})+H(^2S)}$ and $\mathrm{Si^+(^2P_{\frac{1}{2}})+H(^2S)}$ dissociation limits, which are separated by $287\ \mathrm{cm^{-1}}$. The assignments were made by combining experimental information with predictions of the vibrational and rotational energy levels obtained by numerical solution of the radial Schrödinger equation and calculations of the rotational line intensities. The experimental data were obtained by measurement of the transition frequencies, line intensities, linewidths and hyperfine splittings and through examination of the effect of laser power on the linewidth. The kinetic energy released on dissociation and the photofragment angular distribution were also determined.

Introduction

The rovibronic energy levels of well-separated electronic states in a diatomic molecule near to its equilibrium internuclear configuration are usually well described within the Born–Oppenheimer approximation. By using spectroscopic data or the results of *ab initio* calculations, it is possible to obtain potential-energy surfaces that describe the vibrational and rotational motion of the nuclei in this region. At the other extreme, the electronic structures of the isolated atomic fragments can usually be established by electronic spectroscopy and at least an approximate long-range potential between the atoms may be calculated. The last few vibrational levels of a molecule just below dissociation have a different character from those near the bottom of the well (Stwalley 1978), and in the absence of non-adiabatic effects their energies and wavefunctions are almost totally determined by the long-range form of the interatomic potential (Stwalley 1978; LeRoy & Bernstein 1970). The region of intermediate internuclear separation is difficult to probe experimentally and only in a very few cases has the structure of a molecule been investigated up to its dissociation limit.

The near-dissociation behaviour of a diatomic molecule is of particular interest when one or both of the atomic fragments possess electronic angular momentum. It then follows that more than one molecular electronic state must correlate to the same dissociation limit. For example, the four lowest-lying electronic states of the SiH+ molecular ion all correlate with the $\mathrm{Si^+(^2P)+H(^2S)}$ dissociation asymptote as shown in figure 1. In the near-dissociation region, the energy separation of the electronic states becomes comparable to the magnitude of non-

[159]

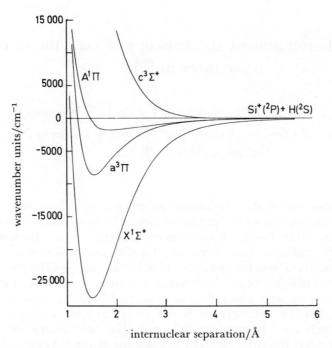

FIGURE 1. Composite experimental–theoretical potential-energy curves for SiH⁺ excluding spin–orbit coupling. The derivation of the singlet curves is described in the text. The ³Π and ³Σ⁺ curves are from Hirst (1986) with the ³Π well located in accordance with the results of Berkowitz et al. (1987).

adiabatic interactions between the Born–Oppenheimer states. Accordingly, it is anticipated that spin–orbit and Coriolis coupling may no longer be treated as a small perturbation within a Born–Oppenheimer basis. The effect of non-adiabatic interactions is expected to be revealed through their influence on the positions of the energy levels just below and also just above the dissociation threshold and also on the character of their corresponding wavefunctions. The molecular fragmentation dynamics will also be very sensitive to these interactions. Levels above dissociation are also quite properly described as scattering resonances. The nature of near-threshold resonances when non-adiabatic interactions are important has recently been examined in a series of theoretical papers by Freed & co-workers, who have predicted the existence of many interesting novel effects in the photofragmentation spectra of diatomic molecules that dissociate into open-shell atoms (Singer et al. 1985; Williams & Freed 1986; Williams et al. 1987).

In principle, a wealth of information on the intermediate (recoupling) region may be obtained from spectroscopic experiments on energy levels just below threshold. In practice, it is probably easier to obtain spectra involving energy levels just above dissociation by recording photofragment spectra. The spectroscopic experiment allows detection of much narrower resonances than is possible in atomic scattering experiments that are subject to relatively poor translational-energy resolution and no control over the impact parameter. One experimental approach to the study of near-threshold resonances is via the laser photodissociation of molecular ions in fast ion beams (for a review, see Moseley 1985), and in this paper we describe experiments of this type on the SiH⁺ ion. Feshbach resonances are found to be dominant and the first evidence for multichannel resonances in the photodissociation spectrum of a diatomic molecule is described.

NEAR-THRESHOLD RESONANCES IN THE DISSOCIATION OF DIATOMIC MOLECULES

Resonances (or metastable levels) that lie just above the dissociation threshold can be considered to arise in two ways. Firstly, levels may be trapped behind a 'hump' or barrier in the potential. A barrier can arise from the contribution of a centrifugal term to the effective potential because of rotation of the molecule, although a 'hump' may already exist in the potential of the non-rotating molecule because of an avoided crossing between states of the same symmetry. Secondly, Feshbach resonances occur where bound levels that belong to one potential curve lie embedded in the continua of one or more electronic states that correlate to a lower dissociation asymptote. We include in the following discussion those resonances that only occur in the near-threshold dissociation region (for a more general discussion see Lefebvre-Brion & Field 1986).

Shape resonances, which can also be described as centrifugally bound levels, arise from the introduction of a centrifugal barrier in the effective potential of a rotating diatomic molecule. The centrifugal energy term is added to the potential of the non-rotating molecule $U_0(R)$ to yield an effective potential, $U_{eff}(R)$, which may be written

$$U_{eff}(R) = U_0(R) + (\hbar^2/2\mu R^2)\,[J(J+1)-\Omega^2].$$

Solution of the Schrödinger equation for the nuclei with this potential yields eigenenergies that lie below and also above the dissociation asymptote of $U_0(R)$. Those levels that are above the dissociation limit are termed shape resonances and have associated predissociation lifetimes that are determined by the probability of tunnelling through the centrifugal barrier. In scattering terminology such quasibound levels are 'single channel' resonances, as only one potential energy surface is involved. Examples of shape resonances in molecular ions recently observed by fast-ion-beam laser photofragment spectroscopy include studies of HeH^+ (Carrington et al. 1981, 1983) and CH^+ (Helm et al. 1982; Carrington & Softley 1986; Sarre et al. 1986, 1987). The situation is in fact more complex in a molecule such as CH^+ or SiH^+ because shape resonance levels of one state lie in the continua of other electronic states that correlate to the same dissociation limit. With reference to figure 1, it can be seen that a rotationally quasibound level of the $^1\Pi$ state in SiH^+ would be susceptible to predissociation by coupling to the continua of the $^1\Sigma^+$ and $^3\Pi$ states. The influence of these states has been identified in the linewidths recorded in the corresponding photopredissociation spectra of the isovalent molecule CH^+ (Graff et al. 1983; Sarre et al. 1986, 1987). The language employed here assumes that the level structure and predissociation rates can be treated satisfactorily within a Born–Oppenheimer basis. In broad terms this appears to be the case for the high-lying quasibound levels of CH^+ that have been studied to date. A good example of the case where there is already a 'hump' in the potential is found in the absorption $(A^1\Pi - X^1\Sigma^+)$ spectrum of the AlH molecule (which is isoelectronic with SiH^+). The $^1\Pi$ curve has a potential 'hump' (Herzberg & Mundie 1940) with a height of ca. 0.15 eV (Hurley 1961) and it is found that the higher rotational levels of $v' = 0$ and 1 are trapped behind the barrier but lie above the dissociation limit (Herzberg 1950). For the SiH^+ molecule, ab initio calculations show no evidence for a 'hump' in the potential of the $^1\Pi$ state (Hirst 1986, 1987) so this aspect is not of direct importance here.

Secondly, Feshbach resonances occur when an energy level that is nominally derived from one potential energy surface and that is located below the dissociation asymptote to which that surface correlates, lies above a lower dissociation limit or channel. An example is provided in the work reported here and is illustrated in figure 2. The $^1\Pi_1$ state correlates to the upper $Si^+(^2P_{\frac{3}{2}}) + H(^2S)$ fine-structure dissociation limit and so a number of levels of this state lie between the fine-structure dissociation asymptotes. The photodissociation spectra of SiH$^+$

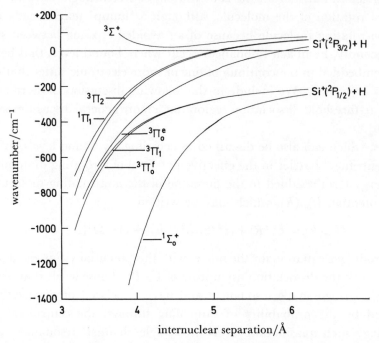

FIGURE 2. Correlation of molecular and atomic states for $J = 2$ including spin–orbit coupling. For simplicity, only seven of the twelve potentials are shown. The spin-orbit splitting in the Si$^+$ ion is 287 cm^{-1}.

recorded in this study arise predominantly from transitions to levels between the two dissociation limits. Predissociation occurs by coupling of the $^1\Pi$ levels to states that correlate to the lower asymptote, i.e. SiH$^+(^2P_{\frac{1}{2}}) + H(^2S)$, which lies 287 cm^{-1} below the Si$^+(^2P_{\frac{3}{2}}) + H(^2S)$ limit. A similar case was found in the infrared predissociation spectrum of HeNe$^+$ (Carrington & Softley 1985).

In reality, this outline of near-threshold resonances is an oversimplification of the problem for SiH$^+$. Our results show evidence of strong mixing of Born–Oppenheimer states through spin-orbit and Coriolis interactions, which have a profound effect on the spectrum. The simplest photodissociation spectrum of SiH$^+$ that could be obtained would arise from excitation from the X$^1\Sigma^+$ state to predissociated levels of the A$^1\Pi$ state for which the selection rules are $\Delta J = 0$ (Q) and $\Delta J = \pm 1$ (R, P). The excited-state levels could be either of the shape resonance or Feshbach type. Non-adiabatic couplings tend to weaken this 'isolated state' model in the near-dissociation region and consequently additional lines may be found in the spectra. This is in accord with theoretical predictions of the photodissociation cross sections based on the results of fully coupled multichannel calculations of the resonances (Williams et al. 1986, 1987). The extra lines involve upper-state levels that are derived from potentials other than the $^1\Pi$ state and that correlate to the Si$^+(^2P) + H(^2S)$ asymptotes. When substantial mixing involving two or more different states occurs, it is appropriate to call the resultant resonances (levels) 'multichannel' in character. For the CH$^+$ molecule, it has been predicted that multichannel

[162]

resonances would be found on photoexcitation from the ground electronic state (Williams *et al.* 1986, 1987), but to date no firm experimental evidence for such resonances has been reported. In fact, an alternative and completely different assignment has been proposed for the known additional lines that occur in the CH^+ photodissociation spectrum near 540 nm (Helm *et al.* 1982; Sarre *et al.* 1986, 1987) and also near 350 nm (Cosby *et al.* 1980; Sarre & Whitham 1987). Complex spectra in the infrared predissociation spectrum of CH^+ have been reported (Carrington & Softley 1986), but the authors did not find any evidence for multichannel resonances. In this paper we describe the observation of multichannel resonances in the photodissociation of SiH^+. To the best of our knowledge this is the first identification of such resonances in the photodissociation spectrum of any molecule.

EXPERIMENTAL CONSIDERATIONS

Laser photofragment spectra are recorded by irradiation of a mass-selected ion beam with tunable laser radiation and detection of the photoproduct ions. The apparatus is illustrated in figure 3. A 10^{-7} A beam of SiH^+ ions is generated by 70 eV electron-impact ionization of silane

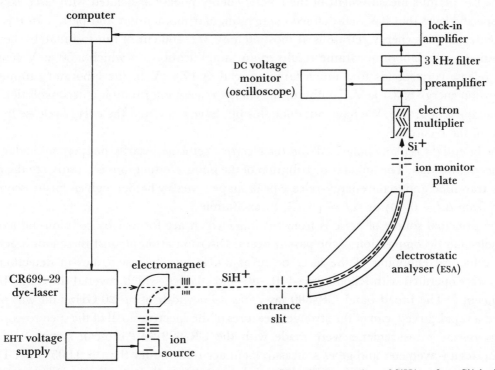

FIGURE 3. Schematic diagram of the fast-ion-beam apparatus. Photofragmentation of SiH^+ to form Si^+ is shown.

in a conventional mass-spectrometric ion source and is accelerated to an energy of 3.7 keV. A small 6 cm radius electromagnet provides mass separation of the SiH^+ ions from the other ions created in the ionization process. Tunable radiation from a Coherent CR-590 (broadband) or CR-699-29 Autoscan (single mode) dye laser is merged coaxially with the ion beam over a path length of 0.5 m and induces transitions from levels of the ground electronic state to near-threshold predissociated levels. Provided that the lifetime of the level towards dissociative decay is sufficiently short, the parent ion dissociates in the flight region into Si^+ and H fragment atoms. An electrostatic energy analyser (ESA) is used to separate the Si^+ ions from the parent beam and these are then detected with an electron multiplier. The ESA also serves to measure the translational energy released in the fragmentation process.

When the single-mode laser is employed, the spectroscopic resolution is determined by the energy spread in the SiH^+ beam. This is typically 0.5 eV (FWHM) and, because of the effect of kinematic compression in an accelerated beam (Kaufman 1976), this corresponds to a Doppler width of only 35 MHz at a laser wavenumber of $17500 \, cm^{-1}$. This narrow linewidth is of considerable value as it is sufficient to permit the resolution of proton nuclear hyperfine splittings in many instances. In other cases, where the predissociation lifetime of the level is sufficiently short, the effect of lifetime broadening outside the Doppler width can be observed. Hyperfine splitting and lifetime broadening are generally mutually exclusive features in the spectra. The laser delivers a power of up to 950 mW and this is sufficient to saturate many of the transitions and induce power broadening. Consequently the power has to be reduced for some of the measurements. Transitions to levels with triplet character have a smaller oscillator strength and in most cases do not exhibit line broadening at high laser power. This is also useful in assigning the spectrum.

The ESA permits measurement of the kinetic-energy release associated with laser excitation to a specific level that has well-defined energy, angular momentum and parity. In this work, the centre-of-mass energy release is at most 50 meV (*ca.* $400 \, cm^{-1}$) but fortunately the transformation to the laboratory frame results in an 'amplification' in which a 50 meV centre-of-mass release corresponds to a much larger spread of 17.4 eV in the laboratory frame for a parent-ion energy of 3.7 keV. To obtain the energy release information a deconvolution of the peak shape is desirable. We have not done this but have estimated the energy release from the FWHM of the recording.

The laser light is plane polarized and the electric vector necessarily lies perpendicular to the ion-beam direction. The angular distribution of the photoproduct ions is sensitive to the nature of the transition and so the energy-release peak shape reveals whether a given spectroscopic line arises from $\Delta J = 0$ (Q) or $\Delta J = \pm 1$ (R, P) excitation.

The principal source of noise is from Si^+ ions, which are formed by collision-induced and unimolecular decomposition of the parent beam. Discrimination against these ions is achieved by mechanical chopping of the laser beam at 3 kHz and by using lock-in detection. The lasers were operated with Rhodamine-110, -6G and -B dyes, which covered the range 18750–$15600 \, cm^{-1}$. The broad-band CR-590 laser has a linewidth of *ca.* 20 GHz and was used to record a rapid survey scan of the strongest features of the spectrum. All of the spectroscopic and energy-release measurements were made with the CR-699-29 Autoscan system, which incorporates a wavemeter and gives a measurement accuracy better than $\pm 0.005 \, cm^{-1}$. The use of a fast beam introduces a Doppler shift of about 8–$10 \, cm^{-1}$ depending on the wavelength. The ion-beam energy and precise Doppler shifts were determined by measuring the frequency of a strong narrow spectroscopic line with both parallel and antiparallel coaxial irradiation of the laser and ion beams.

SPECTROSCOPY AND POTENTIAL–ENERGY CURVES OF SiH^+

The first spectroscopic data on SiH^+ were obtained by Douglas & Lutz (1970), who recorded the 0–0, 0–1, 0–2, 1–0 and 1–1 bands of the $A^1\Pi$–$X^1\Sigma^+$ system in emission from a hollow-cathode discharge containing helium with a trace of silane. This led to the definitive identification of SiH^+ in the solar photosphere (Grevesse & Sauval 1970). The $^1\Pi$ state was found to have a very shallow well with a $\Delta G_{\frac{1}{2}}$ value of only 390.17 cm^{-1}. In subsequent work (Carlson et al. 1980), two additional bands (2–0 and 3–0) were recorded and analysed and the lifetimes for $v' = 0$, 1, 2 and 3 were measured. The observation of transitions involving $v' = 2$ and 3 led to a revision of the dissociation energy of the $^1\Pi$ state and a value for $D_0(^1\Pi)$ of 1230 ± 210 cm^{-1} $(0.15 \pm 0.03$ eV) was deduced.

A photoionization mass-spectrometric study of the SiH radical has provided new information including a measurement of the ionisation potential of SiH (Berkowitz et al. 1987). A number of autoionization features were observed, some of which were attributed to a Rydberg series converging to the $a^3\Pi$ state of SiH^+. This is the first experimental information on the location of this triplet state with respect to the singlet-state manifold. The difference in energy between the $X^1\Sigma^+$ and $a^3\Pi$ states was deduced to be 2.30 ± 0.01 eV, which is in reasonable agreement with the results of two ab initio calculations (Bruna et al. 1983; Hirst 1986).

Ab initio calculations of the SiH^+ potential-energy curves have been reported by Bruna & Peyerimhoff (1983), Bruna et al. (1983) and Hirst (1986), although the first two publications did not contain the numerical data. Hirst (1986) calculated the surfaces for the $X^1\Sigma^+$, $A^1\Pi$, $a^3\Pi$ and $c^3\Sigma^+$ states by using the multireference CI method and showed that there was no evidence for a maximum in the $^1\Pi$ state.

RESULTS AND ASSIGNMENTS

This work was initiated to provide a complementary study to the photodissociation spectrum of CH^+ in which numerous shape resonances have been identified (Helm et al. 1982; Sarre et al. 1986, 1987a). In SiH^+, the $A^1\Pi$ state is substantially shallower than the corresponding state in CH^+ for which $D_0(^1\Pi)$ is 1.159 eV (Helm et al. 1982; Graff et al. 1983). In contrast to CH^+, we have found that the photodissociation spectrum is dominated by Feshbach resonances. A significant difference between the two cases is the larger spin–orbit splitting of 287 cm^{-1} in Si^+ compared with 63 cm^{-1} in C^+. In figure 2 it is shown that the $A^1\Pi$ state correlates to the upper fine-structure limit and so a substantial number of the $^1\Pi$ levels lie between the two asymptotes. Excitation to these levels may then result in fragmentation along the $Si^+(^2P_{\frac{1}{2}}) + H(^2S)$ dissociation channel. We emphasize that the description of such levels as belonging to the $^1\Pi$ state is an approximation but it is a good starting point for the spectroscopic analysis.

The main features in the spectrum were obtained with the CR-590 broad-band laser and the complete spectrum from 15 600 to 18 750 cm^{-1} was then recorded with the CR-699-29 single-mode laser with a data-sampling interval of 75 MHz. Although this is a wider interval than the FWHM of the narrowest lines, it is sufficient to detect all but the weakest transitions. Each line was then examined in turn with a smaller sampling interval. A stick diagram of the spectrum is shown in figure 4 and the wavenumbers of the assigned lines are given in table 1. The quoted accuracy of each measurement varies according to the linewidth and intensity of the line.

The assignments could not have been achieved by using standard techniques such as combination differences etc., alone. The spectrum represents a small subset of all possible A–X transitions involving only those levels that lie just above the dissociation limit and with predissociation lifetimes lying approximately between 10 μs and 1 ps. These lifetime limits are set by the ion flight time in the apparatus and the breadth of the line respectively. Whereas molecular parameters for $v'' = 0$, 1 and 2 and $v' = 0$, 1, 2 and 3 are known from emission

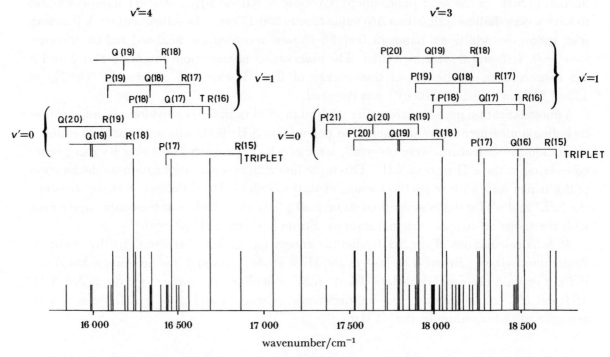

FIGURE 4. Stick diagram of the photofragment spectrum of SiH$^+$ obtained by the detection of Si$^+$ ions. The spectrum was recorded with a data sampling interval of 75 MHz at a scan rate of 5 cm^{-1} min^{-1} and with a 300 ms time constant. The experimental intensities are estimated values only.

TABLE 1. VACUUM WAVENUMBERS OF TRANSITIONS FROM THE $X^1\Sigma^+$ STATE TO EXCITED-STATE PREDISSOCIATED LEVELS (RESONANCES) OF SiH$^+$

(The transitions are to A$^1\Pi$ levels except for a number that have triplet (T) parentage.)

v'	v''	J'	R $(J'-1)$	Q (J')	P $(J'+1)$
0	3	19	18047.32 (1) †	17788.86 (1)	17526.56 (1)
	3	19 T‡	—	17795.51 (1)	—
0	4	19	16234.11 (1)	15983.29 (1)	—
	4	19 T	—	15989.94 (1)	—
0	3	20	17908.7 (1)	17638.60 (1)	17363.6 (1)
0	4	20	16103.2 (1)	15841.06 (1)	—
	3	16 T	18701.51 (1)	18480.69 (1)	18255.73 (1)
	4	16 T	16867.79 (1)	—	16435.30 (1)
1	3	17	18518.91 (1)	18285.44 (1)	18047.85 (1)
	3	17 T	18472.84 (1)	—	18001.77 (1)
1	4	17	16691.64 (1)	16465.00 (1)	16234.65 (1)
	4	17 T	16645.57 (1)	—	—
1	3	18	18388.7 (1)	18145.2 (10)	17892.5 (1)
1	4	18	16568.3 (1)	16331.7 (10)	16087.0 (1)
1	3	19	18246.3 (10)	17989.4 (10)	17725.6 (10)
1	4	19	16433.5 (10)	16183.5 (10)	—
2	3	18	—	18322.1 (2)	—
2	4	18	—	16509.0 (10)	—

† The number in parentheses is the uncertainty in the last quoted figure(s).
‡ Levels labelled T have triplet parentage.

studies, the transitions are found to originate in $v'' = 3$ and 4, and an extrapolation to the levels of interest with Dunham coefficients is not possible. Fortunately, the laser-photofragment and energy-release spectra contain information that would not normally be obtained in an absorption spectrum and, coupled with a different theoretical approach, assignment of a substantial part of the spectrum has been achieved.

The spectrum of figure 4 shows that the lines are mostly clustered in regions near 18000 and 16000 cm^{-1}. The separation of these two regions is approximately the energy-level difference between $v'' = 4$ and $v'' = 3$ of the $X^1\Sigma^+$ state and so many of the lines in these two regions have common upper-state levels. Most of the lines carry a signature that is either a hyperfine splitting or a lifetime-broadened width as listed in table 2 and illustrated in figure 5. The measurement

TABLE 2. WIDTHS (RECIPROCAL CENTIMETRES) AND HYPERFINE SPLITTINGS (MEGAHERTZ) OF
LINES OBSERVED IN THE VISIBLE LASER PHOTODISSOCIATION SPECTRUM OF SiH$^+$

v'	J'	e/f†	linewidth/cm^{-1}	hyperfine splitting/MHz
0	19	e	< 0.0015	120 (10)
0	19	f	< 0.0015	—
	19	fT	< 0.0015	425 (10)
0	20	e	0.6 (1)‡	—
0	20	f	0.05 (1)	—
	16	eT	0.003 (1)	40 (20)
	16	fT	0.002 (1)	235 (10)
1	17	e	0.018 (3)	—
	17	eT	< 0.0015	270 (10)
1	17	f	0.002 (1)	150 (10)
1	18	e	0.5 (1)	—
1	18	f	3.0 (10)	—
1	19	e	2.5 (5)	—
1	19	f	4.5 (10)	—
2	18	f	4.5 (10)	—

† The e/f notation for the Λ-doublet components follows that given by Brown *et al.* (1975). The e and f levels are accessed in $\Delta J = \pm 1$ and $\Delta J = 0$ excitation respectively.
‡ The number in parentheses is the uncertainty in the last quoted figure(s).

of the energy released on fragmentation gives valuable information on the location of the energy level with respect to the dissociation channel, although the precision is very poor in comparison with the spectroscopic data. From the energy-release peak shape it is possible to determine if a transition involves $\Delta J = 0$ or $\Delta J = \pm 1$ excitation. For $\Delta J = 0$ (Q-line), the Si$^+$ fragments are ejected preferentially along the ion beam direction and this leads to a dip in the profile (see figure 6). For $\Delta J = \pm 1$ (R, P-lines), the scattering is perpendicular to the beam direction and a single peak results. The measured releases and profile information is given in table 3.

Following an approach used by Helm *et al.* (1982) in a study of shape resonances in CH$^+$, the assignment of quantum numbers was achieved principally via numerical solution of the Schrödinger equation to calculate the energy levels in the two states. From the emission data (Carlson *et al.* 1980), numerical potentials have been derived for the lower parts of the $X^1\Sigma^+$ and $A^1\Pi$ states (M. Larsson, personal communication 1986). The long-range part of the

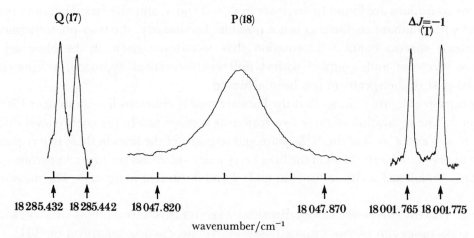

FIGURE 5. Spectroscopic transitions showing hyperfine structure and lifetime broadening for the Q (17) and P (18) lines of the 1–3 band and also for the corresponding $\Delta J = -1$ transition from $v'' = 3$, $J'' = 18$ to a triplet level (T) with $J' = 17$ (e). The hyperfine splitting in the Q (17) line is resolved only when the laser power is *ca.* 30 mW. The line involving the triplet level was recorded with a laser power of over 500 mW and does not exhibit power broadening.

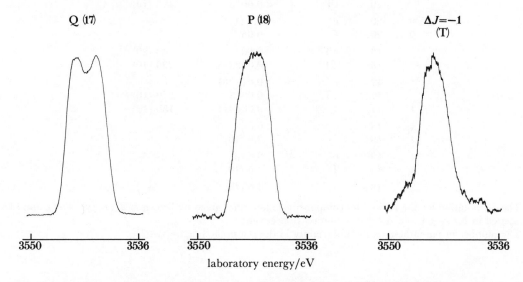

FIGURE 6. Energy-release profiles corresponding to laser excitation of the Q (17) and P (18) lines and to the triplet level of figure 5. The width of the energy release peak for excitation to the triplet $J' = 17$ level is narrower than for the P (18) line in agreement with the spectroscopic shift of -46.1 cm^{-1} (see figure 5 and table 2).

potentials were taken to be determined by the ion-induced dipole term $-C_4/R^4$, where C_4 is a constant involving the polarizability of the hydrogen atom. The two regions were linked smoothly with the shape forced to conform to the expression $-C_4/R^4 - C_6/R^6 - C_8/R^8$, where the C_6 and C_8 coefficients are free parameters. This is a crude approximation and other approaches are being evaluated. However, it is a particularly convenient form when a number of trial potentials with different dissociation energies are being explored. The short-range repulsive potential was linked smoothly using an exponential function. It is also necessary to define the dissociation energies of the states and recognize that the $^1\Sigma^+$ and $^1\Pi$ states correlate to different asymptotes (see figure 2). Initially, the published dissociation energies (Carlson *et al.* 1980)

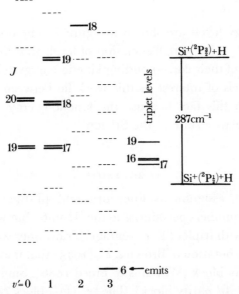

FIGURE 7. Identified singlet and triplet levels of SiH$^+$ close to the dissociation asymptotes. The Λ-doublet splittings are shown schematically and are not drawn to scale.

TABLE 3. CALCULATED AND EXPERIMENTAL 'ENERGY' RELEASES AND PEAK SHAPES ASSOCIATED WITH LASER EXCITATION TO THE RESONANCES LISTED IN TABLE 1

v'	J'	e/f	calculated/cm^{-1}†	experimental/cm^{-1}‡	peak shape§
0	19	e	92	102 (10)	S
0	19	f	92	—	—
	19	fT	99	—	—
0	20	e	203	202 (50)	S
0	20	f	203	—	—
	16	eT	54	55 (15)	S
	16	fT	54	72 (15)	D
1	17	e	92	95 (10)	S
1	17	f	92	107 (10)	D
	17	eT	46	49 (15)	S
1	18	e	191	187 (15)	S
1	18	f	191	215 (25)	—
1	19	e	290	312 (140)	—
2	18	f	363	395 (50)	—

† Calculated level position with respect to the $^2P_{\frac{1}{2}}$ dissociation asymptote.
‡ 'Energy' release estimated from the FWHM of the peak.
§ S and D indicate that the peak shape is a single (S) or double (D) feature. Theoretically, the e and f levels should correspond to S and D, respectively (see text).

were employed but were later found to require upward revision. The bound and rotationally quasibound eigenvalues were determined by numerical solution of the Schrödinger equation for the chosen potentials by using a computer program of R. J. LeRoy (personal communication 1986). The effect of Λ-doubling in the $^1\Pi$ state is not included. We emphasize that these potentials are far from being spectroscopically accurate but this approach is much superior to the attempted extrapolation of molecular parameters to the levels of interest. The rotational line intensities were also calculated by using an *ab initio* electronic transition moment function

(M. Larsson, personal communication 1986) and the same computer program and also were of help in the assignment.

The near-threshold energy levels are shown in figure 7. The observation of emission from $v' = 3$, $J' = 6$ (Carlson et al. 1980), predissociation of levels that lie less than 287 cm^{-1} above this level and measurement of their corresponding kinetic-energy releases (table 3), necessarily means that most of the levels of interest in this work lie between the asymptotes as shown. It is important to establish this fact because the kinetic-energy releases do not carry any information on the fine-structure state of the Si$^+$ ion.

Discussion

Tables 1 and 2 contain the assignments, linewidths and splittings. In many cases it has been possible to assign quantum numbers pertaining to the $^1\Pi$ state, but we also list other lines which involve excited-state levels with triplet (T) parentage. Transitions with $\Delta J = \pm 1$ (R, P) involve excitation to 'e' levels in the notation of Brown et al. (1975) that may be described alternatively as belonging to the 'A' parity block (Williams & Freed 1986). Similarly, the $\Delta J = 0$ (Q) lines involve the 'f' levels (in the 'B' parity block). The 'e' ('A') block contains levels from the $^1\Sigma^+$, $^1\Pi$, $^3\Pi_{2,1,0}$ and $^3\Sigma_1$ states and the 'f' ('B') block has levels from $^1\Pi$, $^3\Pi_{2,1,0}$ and $^3\Sigma_{1,0}$. For the near-threshold levels, the only strictly good quantum numbers are the total angular momentum, J, and parity (neglecting nuclear spin). Interactions between states can only occur with the selection rule $\Delta J = 0$ and within the same parity block. It follows that only one of the components of the $^1\Pi$ state is mixed with the ground state under the action of the Coriolis operator. This causes both the Λ-doublet splitting in the $^1\Pi$ state and also the Coriolis-induced predissociation of the 'e' component of the $^1\Pi$ state via the continuum of the ground electronic state.

The lines exhibit a very wide range of linewidths from the Doppler width of 0.0012 cm^{-1} to ca. 4.5 cm^{-1}. The variation in linewidths between 'e' and 'f' components and also with J value is indicative of erratic couplings of the levels and continua in the near-threshold region. Although it is attractive to attribute the greater linewidth of the 'e' relative to the 'f' component in the levels $v' = 0$, $J' = 20$ and $v' = 1$, $J' = 17$ to the effect of Coriolis coupling with the ground-state continuum, the reverse trend is found for $v' = 1$, $J' = 18$ and 19, so it is clear that additional couplings with other states must be considered. In fact the Q-lines can only appear in the spectrum because of coupling with states of triplet parentage. Subject to the selection rules, all of the levels are subject to couplings with the shape resonances, Feshbach resonances and continua of the other singlet and triplet states in the vicinity. It is likely that a quantitative account for the predissociation linewidths will necessitate fully coupled multichannel calculations.

A consequence of non-adiabatic interactions is the appearance of the extra 'multichannel' resonances in the photodissociation spectrum as predicted theoretically (Singer et al. 1985; Williams & Freed 1986; Williams et al. 1987) and listed in tables 1–3. It is possible to assign the J and parity quantum numbers as indicated because of the spectroscopic line separations from identified singlet–singlet transitions or by extrapolation of ground-state level differences as in the case of the $J' = 16$ levels.

The narrow splittings in table 2 between 40 and 425 MHz are nuclear hyperfine splittings and arise from the presence of the proton nuclear spin of $\frac{1}{2}$. For a pure $^1\Pi$ state, only the electronic orbital–nuclear hyperfine interaction is present but it will be of negligible magnitude for the high-J levels of interest here. For triplet levels or levels with triplet character, the Fermi contact (isotropic hyperfine coupling), spin–spin dipolar, orbital–nuclear and hyperfine-doubling interactions may contribute. The most important is expected to be the Fermi contact term, which provides a direct measure of the electron spin density at the proton nucleus. In the hydrogen atom, the Fermi contact parameter for an electron in the 1s orbital takes the value of 1420 MHz. However, for a molecule with a pure $^3\Pi$ state arising from a $\sigma^1\pi^1$ configuration, only half of the spin angular momentum contributes significantly to the spin density at the proton nucleus so the expected maximum observed splitting will be $1420/2 = 710$ MHz. All of the experimental splittings are indeed lower than this value. The existence of a splitting gives a qualitative indication of the triplet character in the wavefunction associated with a given resonance, although the absence of a splitting does not imply that the line involves a singlet level. Table 2 shows that even some of the levels which are nominally assigned to the $^1\Pi$ state can show significant splittings. The levels $v' = 0$, $J' = 19$ (e) and $v' = 1$, $J' = 17$ (f) exhibit splittings of 120 MHz and 150 MHz respectively because of the admixture of triplet character.

CONCLUSIONS

We have recorded and assigned a substantial part of the visible laser photodissociation spectrum of the SiH^+ ion and have established the principal features of the energy level (resonance) structure in the near-threshold region. The importance of non-adiabatic inter-actions is shown through the irregular behaviour of the linewidths, nuclear hyperfine splittings and the appearance of extra lines. Multichannel resonances are positively identified, we believe for the first time.

Ultimately, it is important to establish quantitative agreement between theory and the observed resonance energies, widths and hyperfine splittings. It seems improbable that a perturbation treatment will be able to achieve this objective and we are currently performing fully coupled calculations of the level (resonance) structure in an effort to achieve this objective.

We are most grateful to Dr D. M. Hirst for calculating the *ab initio* potential curves and for his interest in this work. We thank Professor R. J. LeRoy for a copy of the program LEVEL and Dr M. Larsson for sending the results of his work before publication. We have enjoyed many stimulating conversations with Professor K. F. Freed and Dr C. J. Williams.

We thank the SERC and the Research Corporation Trust for grants towards the purchase of apparatus and the SERC for a studentship to J.M.W.; P.J.S. thanks the Nuffield Foundation for the award of a Science Research Fellowship and C.J.W. thanks the National Westminster Bank for support of a University Research Studentship.

REFERENCES

Berkowitz, J., Green, J. P., Cho, H. & Ruščić, B. 1987 *J. chem. Phys.* **86**, 1235–1248.
Brown, J. M., Hougen, J. T., Huber, K.-P., Johns, J. W. C., Kopp, I., Lefebvre-Brion, H., Merer, A. J., Ramsay, D. A., Rostas, J. & Zare, R. N. 1975 *J. molec. Spectrosc.* **55**, 500–503.
Bruna, P. J. & Peyerimhoff, S. D. 1983 *Bull. Soc. chim. Belg.* **92**, 525–546.
Bruna, P. J., Hirsch, G., Buenker, R. J. & Peyerimhoff, S. D. 1983 In *Molecular ions – geometric and electronic structures* (ed. J. Berkowitz & K.-O. Groeneveld), p. 325. New York: Plenum.
Carlson, T. A., Copley, J., Durić, N., Elander, N., Erman, P., Larsson, M. & Lyyra, M. 1980 *Astron. Astrophys.* **83**, 238–244.
Carrington, A., Buttenshaw, J. A., Kennedy, R. A. & Softley, T. P. 1981 *Molec. Phys.* **44**, 1233–1237.

Carrington, A., Kennedy, R. A., Softley, T. P., Fournier, P. G. & Richard, E. G. 1983 *Chem. Phys.* **81**, 251–261.
Carrington, A. & Softley, T. P. 1985 *Chem. Phys.* **92**, 199–219.
Carrington, A. & Softley, T. P. 1986 *Chem. Phys.* **106**, 315–338.
Cosby, P. C., Helm, H. & Moseley, J. T. 1980 *Astrophys. J.* **235**, 52–56.
Douglas, A. E. & Lutz, B. L. 1970 *Can. J. Phys.* **48**, 248–253.
Graff, M. M., Moseley, J. T., Durup, J. & Roueff, E. 1983 *J. chem. Phys.* **78**, 2355–2362.
Grevesse, N. & Sauval, A. J. 1970 *Astron. Astrophys.* **9**, 232–238.
Helm, H., Cosby, P. C., Graff, M. M. & Moseley, J. T. 1982 *Phys. Rev.* A**25**, 304–321.
Herzberg, G. & Mundie, L. G. 1940 *J. chem. Phys.* **8**, 263–273.
Herzberg, G. 1950 In *Molecular structure and molecular spectra I, spectra of diatomic molecules*, p. 425. New York: van Nostrand.
Hirst, D. M. 1986 *Chem. Phys. Lett.* **128**, 504–506.
Hirst, D. M. 1987 *J. chem. Soc. Faraday Trans.* II **83**, 61–68.
Hurley, A. C. 1961 *Proc. R. Soc. Lond.* A**261**, 237–245.
Kaufman, S. L. 1976 *Optics Commun.* **17**, 309–312.
Lefebvre-Brion, H. & Field, R. W. 1986 In *Perturbations in the spectra of diatomic molecules*. London: Academic Press.
LeRoy, R. J. & Bernstein, R. H. 1970 *J. chem. Phys.* **52**, 3869–3879.
Moseley, J. T. 1985 *Adv. chem. Phys.* **60**, 245–298.
Sarre, P. J., Walmsley, J. M. & Whitham, C. J. 1986 *J. chem. Soc. Faraday Trans.* II **82**, 1243–1255.
Sarre, P. J., Walmsley, J. M. & Whitham, C. J. 1987 *Faraday Discuss. chem. Soc.* **82**, 67–78.
Sarre, P. J. & Whitham, C. J. 1987 (In preparation.)
Singer, S. J., Freed, K. F. & Band, Y. B. 1985 *Adv. chem. Phys.* **61**, 1–113.
Stwalley, W. C. 1978 *Contemp. Phys.* **19**, 65–80.
Williams, C. J. & Freed, K. F. 1986 *J. chem. Phys.* **85**, 2699–2717 and *Chem. Phys. Lett.* **127**, 360–366.
Williams, C. J., Freed, K. F., Singer, S. J. & Band, Y. B. 1987 *Faraday Discuss. chem. Soc.* **82**, 51–66.

Discussion

G. DUXBURY (*Department of Physics and Applied Physics, University of Strathclyde, Glasgow, U.K.*) Dr Sarre and his group have carried out some elegant experiments to show the many resonances that can occur when highly excited states of the diatomic ions SiH^+ and CH^+ interact with continuum states.

In their work on the polyatomic ions, SiH_2^+ and PH_2^+ more complex behaviour is observed. Much of the $\tilde{A}\,^2B_1$–$\tilde{X}\,^2A_1$ spectrum of SiH_2^+ can be understood within the framework of Renner–Teller coupling (Curtis *et al.* 1985). However, in part of the SiH_2^+ spectrum and in the spectrum of PH_2^+ detected via P^+ or PH^+ the explanation is not as clear cut (Jackson 1985). One possible explanation of the spectra involves an extension of the system of diatomic resonance to the polyatomic case, the other is outlined below.

Some recent experiments by Reutt *et al.* (1986) shed some light on possible processes occurring. They carried out molecular-beam photoelectron spectroscopy of both H_2O^+ and D_2O^+, and have interpreted their results as from two principal causes, Renner–Teller coupling between the $\tilde{A}\,^2A_1$ and $\tilde{X}\,^2B_1$ states, and curve crossing between the $\tilde{B}\,^2B_2$ and $\tilde{A}\,^2A_1$ states. This interpretation is based on an autocorrelation function analysis of the high-resolution photoelectron spectra of cold H_2O^+ and D_2O^+. Even though the levels of the $^2B_2$ state are very highly excited, the possibility of multiple intersection between potential-energy surfaces, particularly the conical intersection between the $^2B_2$ and $^2A_1$ surface, is held to be the principal route to rapid decay of the excited state, rather than coupling to the continuum. The possibilities of a much wider range of surface crossings in polyatomic species make the classification of interaction with the continuum states much more complicated than in diatomic molecules.

References

Curtis, M. C., Jackson, P. A., Sarre, P. J. & Whitham, C. J. 1985 *Molec. Phys.* **56**, 485–488.
Jackson, P. A. 1985 Ph.D. thesis, University of Nottingham, U.K.
Reutt, J. E., Wang, L. S., Lee, Y. T. & Shirley, D. A. 1986 *J. chem. Phys.* **85**, 6928–6939.

Phil. Trans. R. Soc. Lond. A **324**, 247–255 (1988)

Printed in Great Britain

Towards a spectroscopy of doubly charged ions

By J. H. D. Eland[1], S. D. Price[1], J. C. Cheney[1], P. Lablanquie[2], I. Nenner[2]
and P. G. Fournier[3]

[1] *Physical Chemistry Laboratory, University of Oxford, South Parks Road, Oxford OX1 3QZ, U.K.*
[2] *LURE, Bâtiment 209c Université Paris Sud, 91405 Orsay, France*
[3] *Spectroscopie de Translation, Bâtiment 478 Université Paris Sud, 91405 Orsay, France*

In the light of a recent demonstration that a two-electron symmetry rule governs final-state selection in the double photoionization of argon, we re-examine current interpretations of molecular doubly charged ion spectra. New double-ionization data from photoionization and charge-exchange experiments on O_2 and C_2H_2 augment the number of centrosymmetric molecules on which the application of the selection rule can be tested. Results do not contradict a rule that photoionization of closed-shell molecules populates chiefly triplet gerade or singlet ungerade states of doubly charged ions.

1. Introduction

Molecules that lose two electrons as a result of an energetic encounter remain as doubly charged ions only under special circumstances. Because all are highly reactive as powerful oxidizing agents they can be kept only either in a vacuum or in a superacid medium. Even in such favourable environments, many doubly charged ions have very short lifetimes, and very few support bound electronically excited states. Because of these characteristics, spectroscopic techniques that rely on the existence of long-lived species can be applied to only a limited subclass of doubly charged ions. Optical emission is known for only three species, N_2^{2+} (Carroll 1958), NO^{2+} (Besnard *et al.* 1986) and O_2^{2+} (Tohji *et al.* 1986). To explore the term schemes of the majority of doubly charged molecular ions it is therefore necessary to measure the energies needed to form them from other well-defined species. They are formed directly from ground-state neutral molecules by photoionization and double-charge transfer, from an ionic core-hole state in Auger spectroscopy and from a normal singly charged ion in charge-stripping spectroscopy. Each of these techniques has its special characteristics. Double-charge transfer to H^+ strictly conserves spin and so populates singlet states exclusively for closed-shell light-atom species (Appell *et al.* 1973; Moore 1973, 1974), but seems, according to recent results, to populate almost all available singlet states (Millié *et al.* 1986). Within the multiplets arising from single electron configurations, Auger transitions also populate singlet states preferentially, although triplets are not entirely excluded (Thomas & Weightman 1981), whereas charge stripping (Porter *et al.* 1981) presumably populates both singlets and triplets equally. The characteristics of double ionization by single photons, which are the subject of this paper, are still not understood. It has become clear that photoionization populates at least ground triplet states of several doubly charged ions (CS_2^{2+}, CO_2^{2+}, H_2O^{2+}, CH_4^{2+}) and the singlet ground state of one (NH_3^{2+}) (Eland *et al.* 1986), but no paradigm has hitherto emerged.

In this paper we set out to examine the systematics of photo-double ionization of molecules with the help of new experimental data, and in the light of a recent demonstration that a strong selection rule operates in the double photoionization of atoms (Lablanquie *et al.* 1987).

2. EXPERIMENTAL PRINCIPLES

Double ionization can be studied in the inevitable presence of single ionization either by observation of a stable doubly charged product, by recognition of the removal of two electrons, or by detection of two singly charged ions in coincidence. In double-charge-transfer spectroscopy (Appell *et al.* 1973), two-electron exchange is selected by using H^+ projectiles and detecting H^- specifically, from the process

$$H^+ + M \rightarrow H^- + M^{2+}.$$

The energy required to effect the charge transfer is abstracted from the translational energy of the proton, and is determined from the kinetic energy of the H^- ion measured at a very small scattering angle. Overall two-electron transfer can also occur in two sequential stages, each of which produces singly charged ion, but such interfering processes are easily recognized and eliminated by their quadratic pressure dependence. Their energy requirements can be accurately predicted from the photoelectron spectrum of M, and provide a useful energy calibration.

Double ionization by photon impact can sometimes be recognized by observation of a doubly charged ion in the mass spectrometer; normally it is necessary to detect, in coincidence, two particles of the same sign from the reaction

$$M + h\nu \rightarrow M_1^+ + M_2^+ + 2e^-.$$

In photoion–photoion coincidence (PIPICO) spectroscopy (Dujardin *et al.* 1984; Curtis & Eland 1985), ionization is brought about in a strong uniform electric field, which accelerates ions into the drift tube of a time-of flight mass spectrometer. The arrivals of correlated ion pairs are recorded as a function of the time difference between them, which is centred on a time difference proportional to the difference in square roots of the ion masses. The arrival-time difference has a dispersion caused by the kinetic energy released in formation of the ion pair, as in normal time-of-flight mass spectrometry (Franklin *et al.* 1967), from which the magnitude of the energy release can be deduced. Multiple energy releases can sometimes be recognized from the shape of the peaks; this can be done reliably, and energy-release distributions can be determined, by restricting detection to those pairs that dissociate along the mass-spectrometer axis (Richardson *et al.* 1986), although this technique is very costly in experimental run time. Once the kinetic-energy release is known it can be added to the energy of the (assumed) dissociation products to provide an estimate of the energy of the precursor doubly charged ion. The yield of the correlated pairs can also be measured as a function of wavelength and the threshold energy for each dissociative double ionization reaction can be determined.

A more direct determination of the doubly charged ion energy can be made by a photo-electron–photoelectron coincidence (PEPECO) experiment, in which two energy-analysed electrons are detected in coincidence. The energy balance is completely defined by

$$h\nu = A(m^{2+}) + E_1 + E_2,$$

and is exactly analogous to the energy balance in photoelectron spectroscopy. Because of the low overall collection efficiency for energy-resolved electrons this is a very difficult experiment, however, and special strategies are needed to make it practicable. A new apparatus at Oxford uses two large hemispherical analysers, which will analyse electrons after acceleration in special lenses designed to accept electrons from a very large solid angle. An alternative strategy, which has enabled the first successful PEPECO experiments at Orsay (Lablanquie *et al.* 1987) is to selectively detect one electron of near-zero kinetic energy, and energy analyse the other. If ionization takes place in a weak electric field, all zero-energy electrons are accelerated parallel to the field lines, and can be detected with very high efficiency in a restricted solid angle. Faster electrons that happen to be ejected into the acceptance angle can be rejected after detection, because their flight times, measured from the time of a light pulse, deviate from the expected value (Morin *et al.* 1980). The efficiency of this 'threshold' electron detection is so high that even with a relatively inefficient 127° cylindrical condenser analyser for the second electron, the electron–electron coincidence experiment proved possible. The interpretation of results from experiments with this configuration does rely on the theoretical prediction (Read 1985), which was, however, directly confirmed, that the electrons ejected in direct double ionization near threshold have a flat distribution.

Experimental details of the double-charge-transfer method (Fournier *et al.* 1985) and of the photoionization techniques (Lablanquie *et al.* 1985) have been given elsewhere, and are not repeated here.

3. Results

This section begins with a review of relevant previous results from PIPICO, double-charge-transfer experiments and the first PEPECO experiment, after which new data on O_2 and C_2H_2 double ionization are presented.

3.1. *Existing data*

The PIPICO technique has been in use for over three years, and detailed studies of ten molecules have been published, with less extensive results on some dozen others (for a review, see Eland *et al.* 1986). In every case of dissociative double ionization where the precursor doubly charged ion energy could be estimated both by addition of a measured kinetic-energy release to the lowest dissociation asymptote and by direct determination of the reaction threshold, the two estimates are in sufficiently close agreement to confirm that ground-state products are formed. It is noticeable that this is true even in cases where the Wigner–Witmer correlation rules predict otherwise (H_2O^{2+}, Richardson *et al.* 1986; CH_4^{2+}, Fournier *et al.* 1985). Some internal excitation of the products is certainly involved, and the uncertainties and spread of kinetic-energy releases still leave the possibility of electronic excitation open in some cases (CO_2, Dujardin & Winkoun 1985; NO_2, Eland *et al.* 1986). Nevertheless, the usefulness of the kinetic-energy measurement as an estimator of doubly charged ion-state energies is confirmed. In addition to the magnitude of the energy release, the width of the Franck–Condon zone can be estimated from the spread of kinetic energies, and the partial cross section for formation of each state is given by the abundance of products with a particular energy. Thus although the PIPICO technique is indirect, it leads to a view of the spectrum of a doubly charged ion containing, in principle, all the information that would be given by a direct photoionization method. As an example, a simulated vertical photoionization spectrum of CS_2 is shown in figure 1, where it is compared with a double-charge-transfer spectrum.

Double-charge transfer to fast H^+ ions seems to populate all, or at least most, singlet states of a doubly charged ion. The relative intensities are distorted by a strong fall-off of the cross

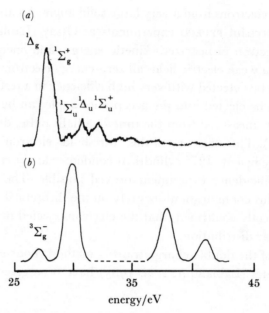

FIGURE 1. Double-charge-transfer (DCT) spectrum (a) of CS_2 at an H^+ energy of 2 keV, and a simulated double photoionization (PI) spectrum (b) for a photon energy of 50 eV.

section at higher energies, for which no compensation can be made. The fall-off is less rapid at higher projectile energies, but at the cost of diminished resolution.

Figure 1 exemplifies the fact that rather few states are populated by double photoionization. They are not the same states as seen in double-charge transer, nor certainly all the accessible states as calculated by *ab initio* theory. The state identifications already marked on figure 1 have been arrived at by comparison with extensive SCF-CI (self-consistent field–configuration inter-action) calculations (Millié *et al.* 1986). According to the calculations the second, and most intensely populated, state in photoionization must be one or more out of $^1\Sigma_u^-$, $^3\Sigma_u^-$, $^3\Delta_u$ or $^3\Sigma_u^+$ arising from the dominant configuration $\pi_u^3 \pi_g^3$. Until now we have had no means of deciding which of these states is involved.

A hope of understanding the selectivity of double photoionization is offered by the results of the PEPECO experiment on argon by Lablanquie *et al.* (1987). They discovered that of the three states ^3P_g, ^1D_g and ^1S_g of Ar^{2+} arising from the p^4 configuration only ^3P_g is significantly populated by photoionization at a photon energy of 56 eV. This observation, which was not expected by the experimenters, actually confirms the theoretical predictions of several groups (Greene & Rau 1982, 1983; Stauffer 1982; Read 1985) in connection with Wannier theory that only certain symmetries of the two-electron wavefunction are favourable to two-electron escape from a positively charged atom. The optimum configuration, called the Wannier point, requires the electrons to remain equidistant from that ion with equal radial velocities at 180° to each other. Unless the three quantum numbers L, (total orbital angular momentum), S (total spin) and Π (parity) are all even or all odd, the two-electron wavefunction has a node at the Wannier point and double ionization is disfavoured. When the dipole selection rule is applied to double photoionization of argon, it is found that only the ^3P_g state of Ar^{2+} can be formed together with one of the favourable electron wavefunction symmetries ^1S_g, ^3P_u, ^1D_g and so on. We now attempt to generalize this rule to molecules.

For photoionization of a closed-shell molecule possessing a centre of inversion, the atomic-symmetry requirement implies a simple rule that states of the double-charged ion formed will be of even (g) parity if spin triplets, or of odd (u) parity if singlets. The reduction in symmetry seems to remove all force from the orbital angular-momentum restriction. The rule can be applied to the CS_2^{2+} spectrum in figure 1, and eliminates $^3\Sigma_u^-$, $^3\Delta_u$ and $^3\Sigma_u^+$ as likely identifications for the second peak, leaving only $^1\Sigma_u^-$. The formation of the ground state of CS_2^{2+}, $^3\Sigma_g^-$, which is observed, is also allowed according to the rule.

If the state near 30 eV populated by photoionization is indeed $^1\Sigma_u^-$, we have three independent estimates of its energy. From the $CS^+ + S^+$ threshold 30.2 ± 0.4 eV, from the kinetic-energy release in PIPICO 30.7 ± 1 eV and from the double-charge transfer spectrum 30 ± 0.3 eV, which are in very good agreement.

Similar reasoning applied to present evidence on CO_2 double photoionization leads to a similar conclusion. It is certain that the ground state of CO_2^{2+}, $^3\Sigma_g^-$, is populated by photoionization, and the next strong population appears at 39.7 ± 0.5 eV as the threshold for $CO^+ + O^+$ formation, close to a feature at 40.6 ± 0.3 eV seen in double-charge transfer. From the estimated accuracy of the energy normalized SCF-CI calculations (Millié *et al.* 1986) the state involved must be either $^1\Sigma_u^-$ from the $\pi_u^3 \pi_g^3$ configuration or $^1\Sigma_g^+$ from $\pi_u^4 \pi_g^2$. Application of the selection rule evidently favours the identification as $^1\Sigma_u^-$, contrary to previous conclusions.

Few other centrosymmetric molecules have yet been examined by double-photoionization techniques in sufficient detail for any test to be made. We have therefore re-examined the PIPICO spectra of oxygen and ethyne (acetylene).

3.2. *Oxygen*

An early PIPICO study of oxygen (Curtis & Eland 1985) showed several distinct energy releases in $O^+ + O^+$ formation after He II ionization. Because of the large number of available states of O_2^{2+} calculated by Beebe *et al.* (1976) and uncertainty as to whether their dissociations are homogeneous, no definite interpretation could be arrived at. Several energy releases were also found after electron impact ionization of O_2 by Brehm & de Frênes (1978) and by Curtis & Boyd (1984) who also studied collision-induced dissociation of O_2^{2+}. There is only partial agreement between the different sets of energy releases, and the PIPICO data seem to be unique in showing energy releases less than 6 eV.

To clarify this we have remeasured the PIPICO spectrum of O_2 both under He II conditions with good resolution for kinetic energies, and also at three selected wavelengths. The results are presented in figure 2. At the longest wavelength, 326 Å† (38 eV) there is a single energy release of 4.5 ± 0.2 eV, which can only arise from ground-state O_2^{2+} ($^1\Sigma_g^+$) dissociating to ground-state products, $2O^+$ (^4S). Because some O_2^{2+} ions formed by photon impact and electron impact are known from mass spectrometry to have lifetimes at least as long as a few microseconds, it seems that the rapid dissociation responsible for the PIPICO signal must come from excitation to part of the O_2^{2+} ($^1\Sigma_g^+$) state outside the potential well. At 304 Å (40.8 eV) we find two energy releases; the 4.5 eV release is still present, but a release of 7.2 ± 0.5 eV gives more abundant $O^+ + O^+$ pairs. According to the calculation of Beebe *et al.* (1976), the state that is most likely to be formed by a vertical transition at this photon energy is $^3\Sigma_u^+$, which is also the only other state (among singlets and triplets) that correlates to ground-state products. The observation positively confirms that this state is formed, because any state correlating with excited products but releasing 7.2 eV would have to lie above $7.2 + 35.7 = 42.9$ eV, and thus above the actual

† 1 Å $= 10^{-10}$ m $= 10^{-1}$ nm.

[177]

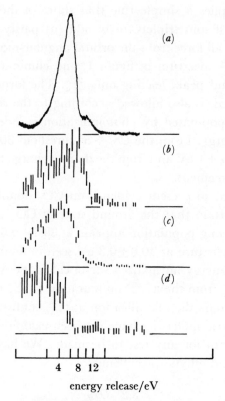

energy release/eV

FIGURE 2. PIPICO spectra for dissociative double ionization of O_2 ($O_2^{2+} \rightarrow O^+ + O^+$) at the wavelengths (b), 25.6 nm; (c) 30.4 nm; and (d) 32.6 nm. The uppermost spectrum, (a), taken with filtered HeII light, is a differential spectrum, whereas the other spectra are of integral form.

photon energy. These PIPICO observations quite definitely locate Franck–Condon zones of the first two states of O_2^{2+}. At the higher photon energy of 48 eV (256 Å) the energy-release distribution closely resembles the one observed with filtered HeII light, including at least three energy releases including 4.5 eV, 7.2 eV and one or more energies near 9.5 eV. The theoretical calculations indicate that more than twelve states of O_2^{2+} are accessible at this energy, but only a few correlate to such low-energy ion-pair products as to release more than 7 eV of kinetic energy.

From the PIPICO results we deduce that part of the Franck–Condon region of the $^1\Sigma_g^+$ ground state of O_2^{2+} is at 36.8 ± 0.2 eV, which is in satisfactory agreement with the electron-impact appearance potentials of O_2^{2+} reviewed by Curtis & Boyd (1984). We locate the $a^3\Sigma_u^+$ state at 39.5 ± 0.5 eV; this may be the state that produces the first Auger peak, at 38.3 eV (Moddeman et al. 1971) and may be the terminus of optical emission (Tohji et al. 1986). The double-charge-transfer spectrum, remeasured recently at high resolution, contains resolved peaks at 41.1 and 43.3 eV, but not at 36.8 or 39.5 eV. Because the excitation of both $X^1\Sigma_g^+$ and $a^3\Sigma_u^-$ is forbidden in double-charge transfer (Appell et al. 1973), this is in agreement with the present assignment. The states at 41.1 and 43.3 eV are probably higher triplet states, whose formation by photon impact may also be possible at short wavelengths.

Because the ground state of O_2 is $^3\Sigma_g^-$, the simple symmetry rule for double photoionization does not apply. Instead we require the two-electron wavefunction symmetry to be one of ^1S_g, ^3P_u or ^1D_g, and then consider which O_2^{2+} states when combined with these symmetries give rise to the dipole-allowed $^3\Sigma_u^-$ or $^3\Pi_u$ overall states. The conclusion is that both $^1\Sigma_g^+$ and $^3\Sigma_u^+$, whose formation is proven, are indeed allowed. Many of the higher-energy states are also allowed, and only $^1\Sigma_u^-$, $^1\Delta_u$ and $^1\Pi_u$ are forbidden, so the identity of the states producing the highest-energy O^+ ions cannot yet be ascertained.

3.3. *Ethyne*

The PIPICO spectra of ethyne and ethyne-d_2 have been examined by using filtered He II radiation (Eland *et al.* 1986), but have not previously been analysed for the energies of the $C_2H_2^{2+}$ states. The intense two-body reactions forming $C_2H^+ + H^+$ and $CH^+ + CH^+$ from C_2H_2 both release kinetic energies which point to a precursor state at 36 ± 0.5 eV (figure 3). The weak three-body dissociations indicate another state at about 41 ± 1 eV. The ground state of $C_2H_2^{2+}$ is $^3\Sigma_g^-$ (Pople *et al.* 1982) and is known to lie near 32.7 eV (Appling *et al.* 1985); at least some, possibly all, ions in this state are stable and therefore do not appear in the PIPICO spectrum. We do, however, clearly observe C_2HD^{2+} ions in the mass spectrum of monodeuteroacetylene under He II photoionization, and consider formation of ground-state doubly charged ions to be the most likely explanation of its presence.

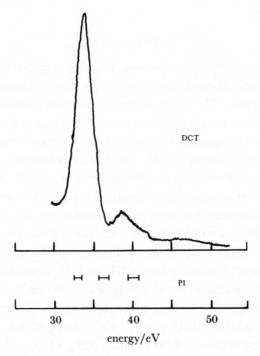

FIGURE 3. DCT spectrum of acetylene, $C_2H_2^{2+}$, and the energies at which photoionization is deduced to occur. High intensity at the low-energy side of the DCT spectrum is caused by interfering two-fold single-ionization processes.

The states of $C_2H_2^{2+}$ accessible by two-electron abstraction from the outermost orbitals of ethyne include only four whose formation would be expected to be favoured according to the symmetry-based rule for photoionization. These are $^3\Sigma_g^-$ (ground state) from $\sigma_u^2 \sigma_g^2 \pi_u^2$, $^1\Pi_u$ from $\sigma_u^2 \sigma_g \pi_u^3$, $^3\Pi_g$ from $\sigma_u \sigma_g^2 \pi_u^3$ and $^1\Sigma_u^+$ from $\sigma_u \sigma_g \pi_u^4$.

To clarify which states are responsible for the dissociative double ionization of ethyne we have re-examined the double-charge-transfer spectrum at higher resolution than before (Appell *et al.* 1974) and have also instituted new scf calculations. The double-charge-transfer spectrum (figure 3) shows $C_2H_2^{2+}$ states, which according to the usual selection rule must be singlets, at 33.7, 38.4, 41 and 45.9 eV. The state at 41 eV is seen only as a weak shoulder in the spectrum; it is within range of the energy of 41 ± 1 eV deduced from the PIPICO results, suggesting that this state may be $^1\Pi_u$, or $^1\Sigma_u^+$. The other states seen in photoionization and double-charge transfer do not coincide, and the supposition must be that the 36 eV state, strongly populated in photoionization, but not by charge transfer is a triplet, perhaps $^3\Pi_g$.

To test the plausibility of this assignment we have carried out scf calculations on $C_2H_2^{2+}$ in a 6-31G** basis. The vertical excitation energies to the first seven states of $C_2H_2^{2+}$ from neutral C_2H_2 are calculated to be 29.3 eV ($^3\Sigma_g^-$), 30.5 eV, ($^1\Delta g$), 31.7 eV ($^1\Sigma_g^+$). 36.5 eV ($^3\Pi_u$), 37.4 eV ($^1\Pi_u$), 38.6 eV ($^3\Pi_g$) and 39.8 eV ($^1\Pi_g$). As these are pure scf calculations without any CI it is not surprising that the lowest state is calculated 3.4 eV lower than the measured appearance potential, and we attempt a first confrontation with experiment by scaling all the energies by $+3.4$ eV. The scaled calculated energy of the $^1\Pi_u$ state is then 40.9 eV, in very good accord with the DCT feature and the PIPICO result of 41 ± 1 eV. The other bands do not match, however, and both $^3\Pi_u$ and $^3\Pi_g$ appear too high to match the PIPICO energy of 36 eV; we are attempting more sophisticated calculations before making a more detailed comparison.

4. DISCUSSION

It is inherently likely that the two-electron wavefunction symmetry, now confirmed to control the direct double photoionization of argon, will also have a strong influence on the direct double ionization of molecules. The effect of such a symmetry restriction can only be to limit the number of dication states strongly populated, in accordance with experimental results. The evidence, which comes from the paucity of distinct kinetic-energy releases in dissociative double ionization, and to a lesser extent from simple straight-line behaviour in double-photoionization cross sections is, however, indirect and could yet be confounded. Many distinct ionic states could release the same kinetic energy by dissociating on parallel surfaces to excited products, and structure in cross-section curves is notoriously unreliable as a guide to ion states. The data on CO_2, CS_2, O_2 and C_2H_2 examined here are consistent with the application of the atomic symmetry rule to these centrosymmetric molecules, but problems still remain. In CO_2^{2+}, for instance, theory predicts two 'allowed' states of $^3\Pi_g$ and $^1\Pi_u$ symmetry near 43 eV, but no sign of them is seen in the photoionization data. Furthermore, the small number of doubly charged ion states in photoionization seems to be a property of centrosymmetric and non-centrosymmetric molecules equally, whereas if the parity rule is dropped the selection rule loses almost all its force. Double photoionization of OCS, SO_2, NH_3, H_2O and CH_4, which are the only molecules yet examined in detail, results, on present evidence, in no more catholic choice of final states than photoionization of centrosymmetric congeners.

Besides the experimental uncertainty, which we have good hope of resolving with the new generation of PEPECO experiments, there is a fundamental question of theory. The electron-pair symmetry rule applies only to direct double photoionization near threshold, and not to Auger-like processes. There is good reason to expect that the selection rules in Auger processes will be different, and it is not yet known whether such processes become important only after inner-shell thresholds, or earlier. As molecular complexity increases, the scope for indirect double ionization involving deep-lying molecular orbitals of the valence shell and inner valence shell must also increase, and it would not be surprising if for large molecules indirect double ionization dominates the direct process even in the range of valence states of the doubly charged ions. At high energies, where inner-shell holes can be created, it is certain that indirect double ionization will be dominant. The role of theory in the advance of this field will be crucial: we now have the means to calculate the energies of molecular and ionic states with really useful accuracy, but the calculation of transition probabilities for double ionization seems to lag behind.

5. Conclusions

Analysis of the present PIPICO and other data is not inconsistent with the idea that the selection rule developed for two-electron escape from atomic ions also applies to molecular double photoionization of small centrosymmetric molecules. If it is confirmed and perhaps extended, the rule will be of great importance in the interpretation of doubly charged ion spectra.

We thank the SERC for financial support.

References

Appell, J., Durup, J., Fehsenfeld, F. C. & Fournier, P. G. 1973 *J. Phys.* B **6**, 197.
Appell, J., Durup, J., Fehsenfeld, F. C. & Fournier, P. G. 1974 *J. Phys.* B **7**, 406.
Appling, J. R., Jones, B. E., Abbey, L. E., Bostwick, D. E. & Moran, T. F. 1985 *Org. Mass. Spectrom.* **18**, 282.
Beebe, N. H. F., Thulstrup, E. W. & Andersen, A. 1976 *J. chem. Phys.* **64**, 2080.
Besnard, M. J., Hellner, L., Malinovich, Y. & Dujardin, G. 1986 *J. chem. Phys.* **85**, 1316.
Brehm, B. & de Frênes, G. 1978 *Int. J. Mass Spectrom. Ion Phys.* **26**, 251.
Carroll, P. K. 1958 *Can. J. Phys.* **36**, 1585.
Curtis, D. M. & Eland, J. H. D. 1985 *Int. J. Mass Spectrom. Ion Phys.* **63**, 241.
Curtis, J. M. & Boyd, R. K. 1984 *J. chem. Phys.* **81**, 2991.
Dujardin, G., Leach, S., Dutuit, O., Guyon, P. M. & Richard-Viard, M. 1984 *Chem. Phys.* **88**, 339.
Dujardin, G. & Winkoun, D. 1985 *J. chem. Phys.* **83**, 6222.
Eland, J. H. D., Wort, F. S., Lablanquie, P. & Nenner, I. 1986 *Z. Phys.* D **4**, 31.
Franklin, J. L., Hierl, P. M. & Whan, D. A. 1967 *J. chem. Phys.* **47**, 3148.
Fournier, P. G., Fournier, J., Salama, F., Richardson, P. J. & Eland, J. H. D. 1985 *J. chem. Phys.* **83**, 241.
Greene, C. H. & Rau, A. R. P. 1982 *Phys. Rev. Lett.* **48**, 533.
Greene, C. H. & Rau, A. R. P. 1983 *J. Phys.* B **16**, 99.
Lablanquie, P., Eland, J. H. D., Nenner, I., Morin, P., Delwiche, J. & Hubin-Franskin, M.-J. 1987 *Phys. Rev. Lett.* **58**, 992.
Lablanquie, P., Nenner, I., Millié, P., Morin, P., Eland, J. H. D., Hubin-Franskin, M.-J. & Delwiche, J. 1985 *J. chem. Phys.* **82**, 2951.
Millié, P., Nenner, I., Archirel, P., Lablanquie, P., Fournier, P. & Eland, J. H. D. 1986 *J. chem. Phys.* **84**, 1259.
Moddeman, W. E., Carlson, T. A., Kranse, M. O. & Pullen, B. P. 1971 *J. chem. Phys.* **55**, 2317.
Moore, J. H. 1973 *Phys. Rev.* A**8**, 2359.
Moore, J. H. 1974 *Phys. Rev.* A**10**, 724.
Morin, P., Nenner, I. Guyon, P. M., Dutuit, O. & Ito, K. 1980 *J. Chim. phys.* **77**, 605.
Pople, J. A., Frische, M. J., Raghavachari, K. & Schleyer, P. v. R. 1982 *J. comput. Chem.* **3**, 468.
Porter, C. J., Proctor, C. J., Ast, T. & Beynon, J. H. 1981 *Croat. chem. Acta* **54**, 407.
Read, F. H. 1985 In *Electron impact ionization*, p. 42. Wien: Springer.
Richardson, P. J., Eland, J. H. D., Fournier, P. G. & Cooper, D. L. 1986 *J. chem. Phys.* **84**, 3189.
Stauffer, A. D. 1982 *Phys. Lett.* A**91**, 114.
Thomas, T. D. & Weightman, P. 1981 *Chem. Phys. Lett.* **81**, 325.
Tohji, K., Hanson, D. M. & Yang, B. X. 1986 *J. chem. Phys.* **85**, 7492.

Phil. Trans. R. Soc. Lond. A **324**, 257–273 (1988)

Printed in Great Britain

257

Formation and destruction of molecular ions in interstellar clouds

By D. Smith

*Department of Space Research, University of Birmingham, P.O. Box 363,
Birmingham B15 2TT, U.K.*

A brief review is given of the role of molecular positive ions in interstellar chemistry. It is indicated how the simplest ions, which are formed by photoionization and cosmic-ray ionization, are converted to polyatomic ions by sequential gas-phase ionic reaction processes such as proton transfer, carbon insertion and radiative association. The importance of H_3^+ and CH_3^+ ions in the initial stages of molecular synthesis and the analogous roles of H_2D^+ and CH_2D^+ (which are formed in the reactions of H_3^+ and CH_3^+ with HD) in the production of deuterated molecules are stressed. Recent work is also mentioned concerning the possible routes to c-C_3H_2 the first cyclic molecule to be detected in interstellar clouds. The process of dissociative recombination of positive ions with electrons is also discussed, because it is commonly invoked as the final step in the destruction of polyatomic ions and in the formation of many of the observed neutral molecules in interstellar clouds, even though the products of such reactions are currently a matter for speculation. It is stressed how spectroscopic studies of the structures and the products of reactions of molecular ions can further advance understanding in the field of interstellar chemistry.

1. Introduction

Several molecular positive-ion species have been detected in interstellar gas clouds (Winnewisser *et al.* 1979; Rydbeck & Hjalmarson 1985; Guelin 1987; Vardya & Tarafdar 1987; Irvine *et al.* 1987). The first was the diatomic ion CH^+, detected in diffuse interstellar clouds via its characteristic absorption bands in the visible region of the spectrum (Dunham & Adams 1937*a, b*; Douglas & Herzberg 1941, 1942). The species HCO^+, N_2H^+, HCS^+ and H_2CN^+, which are all protonated forms of known interstellar molecules, have since been detected in dense interstellar clouds via their characteristic rotational spectra, and very recently SO^+ has also been detected in dense clouds (Churchwell *et al.* 1987). There has also been a tentative identification of the ion HCO_2^+ (Thaddeus *et al.* 1981) and of the COH^+ isomer of HCO^+ (Woods *et al.* 1983). Many other molecular positive-ion species must be present in dense clouds, some in greater concentrations (number densities) than those aforementioned, including ions such as H_3^+, H_3O^+, CH_3^+, NH_4^+ and even more polyatomic species. That some ions have not been detected is simply because of their symmetry, whereas the spectra of others are not yet characterised well enough to allow detection and identification by radioastronomical techniques. Other ionic species will not be present at sufficiently high number densities to be detected until the sensitivity of the detectors associated with radio antennae are increased. It is the combination of advances in laboratory microwave spectroscopy, in ion-structure calculations and in radioastronomy that is resulting in the increase in the detection rate of interstellar molecules (including molecular ions). At the onset, it should be noted that no negative-ion species has yet been detected in interstellar clouds.

The observed interstellar molecular ions and the more numerous interstellar neutral molecules are formed by reactions between simple ions and atoms and molecules in which more complex ions are formed that can react with the electrons present in the weakly ionized interstellar clouds to produce the observed neutral molecules (Dalgarno & Black 1976; Smith & Adams 1981 a, b; Herbst 1985 a; Crutcher & Watson 1985; Adams & Smith 1987 a; Smith 1987). Thus it is the occurrence in the gas phase of many such positive-ion–neutral reactions and positive-ion–electron dissociative recombination reactions that largely constitutes interstellar chemistry, although heterogeneous reactions on the surfaces of interstellar dust grains can also occur (Millar 1982; Duley & Williams 1984; Williams 1987). Indeed, the latter reactions are considered to generate interstellar molecular hydrogen, which is by far the most abundant molecular species (Dalgarno & Black 1976).

Certain key molecular ions are strongly implicated in interstellar gas-phase ion chemistry including H_3^+ and CH_3^+ and in the following sections their central role in the chemistry will be described. Reactions of the ions with H_2 are particularly important. Also the reactions of H_3^+, CH_3^+ and other ions with HD are important in that they result in the creation of deuterated species such as H_2D^+ and CH_2D^+ (Huntress & Anicich 1976; Watson 1976; Smith et al. 1982 a; Smith & Adams 1984 a), the subsequent chemistry of which generates the observed deuterated analogues of several interstellar molecules by the process of isotope exchange and isotope fractionation (Watson 1976; Smith et al. 1982 b; Wootten 1987). Clearly, the spectroscopy of molecular ions has played, and must continue to play, a major role in interstellar physics and chemistry, not least in characterizing the rotational spectra of probable interstellar ions (Winnewisser et al. 1985; Thaddeus et al. 1985; Woods 1987), which assists in their detection and identification, and not least in the elucidation of the neutral products of dissociative recombination reactions that are currently almost totally in the realm of speculation. Some of these important topics will be discussed and highlighted by the examples given in the following sections.

2. PRODUCTION AND REACTIONS OF POSITIVE IONS IN INTERSTELLAR CLOUDS

It is now generally recognized that molecules are synthesized in interstellar clouds largely by the production and reactions of positive ions in the gas phase (Dalgarno & Black 1976; Smith & Adams 1981 a, b; Herbst 1985 a). The understanding of this gas-phase ion chemistry has resulted from astronomical observations of the nature and relative abundances of interstellar molecules (Winnewisser et al. 1979; Guelin 1985; Rydbeck & Hjalmarson 1985; Irvine et al. 1987) (which also provide the vital information on the physical conditions in the clouds), coupled with laboratory gas-phase studies of ionic reactions (Adams & Smith 1983; Smith & Adams 1984 a), which have provided the essential rate data for detailed ion chemical models that aim to describe the overall chemistry (Millar & Freeman 1984 a, b; Leung et al. 1984; Millar & Nejad 1985). These models have had some success in predicting relative molecular abundances that agree with observations. Several reviews are available that describe the ion chemistry of interstellar clouds and that illustrate the successes of the ion chemical models while highlighting some of the remaining problems that ion chemistry has in describing the production of some observed interstellar molecules (Millar 1985; Herbst 1987; Winnewisser & Herbst 1987; Millar et al. 1987). Therefore, it is not the intention here to produce another such review (valuable as they are) but rather to discuss the nature of interstellar ions and their reactivity. In doing this, the vital role of ions in the chemistry will be quite apparent. However, it is pertinent first to briefly describe the composition and the physical conditions pertaining to interstellar clouds.

Interstellar clouds are extremely varied, but they can be categorized into two general types, the so-called diffuse clouds and the dense clouds. The diffuse clouds are of relatively low density consisting mainly of atomic and molecular hydrogen (overall number density of the order of 10^2–10^3 cm^{-3}) and of relatively high temperature (*ca.* 100 K). Their densities are sufficiently low for them to be partly transparent to visible and ultraviolet (uv) radiation (hence the term 'diffuse'). This allows them to be probed via absorption spectroscopy. In this way they are seen to contain elements in the atomic form e.g. H, C, N and O, and simple (diatomic and triatomic) molecules, e.g. CH, CH$^+$, OH, CO, CN and HCN. The ion chemistry occurring is thus relatively simple. It is initiated by the photoionization of carbon atoms by stellar uv radiation producing C$^+$ ions and by the ionization of H and H$_2$ by galactic cosmic rays producing H$^+$ and H$_2^+$ ions. Production of hydrocarbon ions can result from association of C$^+$ with H and H$_2$ terminating at CH$_3^+$. Another vital reaction also occurs that is central also to dense-cloud chemistry:

$$H_2^+ + H_2 \rightarrow H_3^+(v) + H. \tag{1}$$

Reaction (1) is extremely rapid, occurring on every collision between an H$_2^+$ ion and an H$_2$ molecule, and generates the important species H$_3^+$ in a vibrationally excited state (see, for example, Kim *et al.* 1974). The vibrational excitation is removed from the H$_3^+$ in interstellar clouds either by radiation emission (in low-density regions) (Oka 1981) or by energy transfer collisions with H$_2$ molecules (collisional quenching) in denser regions. Relaxed H$_3^+$ ions are unreactive with H$_2$ in bimolecular collisions although at low temperatures in relatively high-pressure gas where termolecular collisions can occur they can associate with H$_2$ to produce H$_5^+$. However, in interstellar clouds, even dense clouds, termolecular collisions are totally improbable and the gas-phase chemistry proceeds only via bimolecular collisions (but, as is discussed later, bimolecular collisions can nevertheless lead to ion–molecule association). Thus the H$_3^+$ ions survive sufficiently long to be able to transfer protons to species such as C and O atoms that have proton affinities greater than that of H$_2$:

$$H_3^+ \underset{O}{\overset{C}{\rightleftharpoons}} \begin{array}{l} CH^+ + H_2 \\ OH^+ + H_2. \end{array} \tag{2a}$$
$$\tag{2b}$$

The CH$^+$ and OH$^+$ product ions of these reactions can then either be photodissociated, or react rapidly with H$_2$ or dissociatively recombine with the ambient electrons regenerating C and O atoms and generating H atoms. Their reactions with H$_2$ generates CH$_2^+$ and H$_2$O$^+$ ions that also react with H$_2$ generating CH$_3^+$ and H$_3$O$^+$ ions (a similar chemistry involving N atoms can also be described). The last-mentioned ions may also be photodissociated (van Dishoeck 1987) or dissociatively recombine with electrons generating small neutral molecules such as the CH and OH radicals observed in diffuse interstellar clouds (Dalgarno & Black 1976). Dissociative recombination and the likely neutral products of such reactions will be discussed more fully in §3. Detailed laboratory studies have shown that protonated ions of stable molecules such as H$_3$O$^+$ are relatively unreactive with H$_2$ even at low temperatures. It is obvious that such ions can therefore survive for relatively long times and may therefore be present in high enough number densities to be detected in interstellar clouds. It is because of this that HCO$^+$, HCS$^+$, N$_2$H$^+$ and H$_2$CN$^+$ are detectable via millimetre-wave emissions from dense clouds.

The gas-phase ion chemistry of dense clouds is much more complex than that of diffuse clouds. As the name implies, the gas-number densities in dense clouds are considerably higher than those in diffuse clouds being typically within the range 10^3–10^6 cm^{-3}. These dense clouds consist largely of molecular hydrogen and atomic helium (presumed to be in their cosmical abundance ratio), and 'dust' grains are dispersed throughout them assisting in rendering them opaque to visible and to shorter-wavelength electromagnetic radiation. Hence these clouds are cooler than diffuse clouds, the temperature being typically 20 K (and perhaps as low as 5 K near to the centre of some dark clouds). These lower temperatures and higher particle-number densities encourage a rich ion chemistry that is initiated by the ionization of H_2 and He by galactic cosmic rays and by which the observed complex molecules in these clouds are synthesized. Such molecules survive against photodissociation because of the shielding afforded by the ambient gas and dust grains. A current list of the molecules observed in interstellar clouds and in circumstellar shells around evolved stars is given in table 1 and includes the seven ion species (in boxes) that have definitely been identified. The most polyatomic species observed to date is $HC_{11}N$, the largest in the family of cyanopolyynes, $HC_n N$ ($n = 3, 5, 7, 9$ and 11). The initial stages of the ion chemistry of dense clouds are quite similar to those outlined above

TABLE 1. MOLECULES DETECTED IN INTERSTELLAR CLOUDS AND CIRCUMSTELLAR SHELLS

hydrogen
H_2

molecules containing only C and H

CH	C≡C	C≡CCH	(C≡C)$_2$H
CH$^+$	C≡CH	(C≡C)$_2$CH	H_3C(C≡C)$_2$H
*CH$_4$	*HC≡CH	H_3CC≡CH	C
	*H_2C=CH$_2$		HC=CH

molecules containing O

OH	HCO$^+$	CH$_3$CHO	CH$_3$CO$_2$H
H_2O	HOC$^+$?	CH$_3$OH	CH$_3$OCH$_3$
CO	H_2CO	CH$_3$CH$_2$OH	HOCO$^+$
HCO	CH$_2$CO	HCO$_2$H	C≡CCO

molecules containing N

CN	H$_2$CN$^+$	CH$_3$CH$_2$CN	H(C≡C)$_4$CN
HCN	NH$_2$CN	H$_2$C=CHCN	H(C≡C)$_5$CN
HNC	CH$_2$NH	HC≡CCN	C≡CCN
N$_2$H$^+$	CH$_3$NH$_2$	H(C≡C)$_2$CN	H$_3$CC≡CCN
NH$_3$	CH$_3$CN	H(C≡C)$_3$CN	H_3C(C≡C)$_2$CN

molecules containing O and N

NO, HNO?, HNCO, HOCN?, NH_2CHO

molecules containing S, Si and Cl

SO, SO$^+$, SN, CS, H_2S, SO$_2$, OCS, HCS$^+$, H_2CS

CH$_3$SH, HNCS, SiO, SiS, *SiC$_2$, *SiH$_4$, HCl

Molecules marked with an asterisk have been detected only in circumstellar shells. The question marks indicate tentative detections. Tentative detections of NaOH (Hollis & Rhodes 1982) and H_2D^+ (Phillips *et al.* 1985) have also been reported. HOCO$^+$ and HOCN are both possible assignments for the same lines (Thaddeus *et al.* 1981). List compiled from various sources including Winnewisser *et al.* (1979), Rydbeck & Hjalmarson (1985), Irvine (1987) and Irvine *et al.* (1987).

for diffuse clouds. H_3^+ is rapidly formed via reaction (1), vibrationally relaxes and then undergoes reactions such as (2) that ultimately produce the observed diatomic molecules (e.g. CO and CS) and small polyatomic molecules (e.g. H_2O, NH_3 and CH_4). Cosmic-ray ionization of He generates He^+, which rapidly reacts with CO producing C^+, which, via reactions with H_2, initiates the growth of hydrocarbon ions. Reactions can then occur between the 'primary' ions C^+ and H_3^+ and various molecules creating other species, e.g.

$$C^+ \xrightarrow{NH_3} H_2CN^+ \xrightarrow{e} HCN. \qquad (3)$$

The H_2CN^+ produced in this reaction undergoes dissociative recombination via which HCN is expected to be produced. Significantly, almost all these small molecules (and certainly the more polyatomic molecules in dense clouds) have proton affinities that exceed that of H_2 (Lias *et al.* 1984) and so they can remove a proton from H_3^+. This is largely the way in which the HCO^+, HCS^+, N_2H^+ and H_2CN^+ observed in dense clouds are produced. Numerous laboratory studies (Bohme 1975; Bohme *et al.* 1980) have shown that when proton transfer is energetically

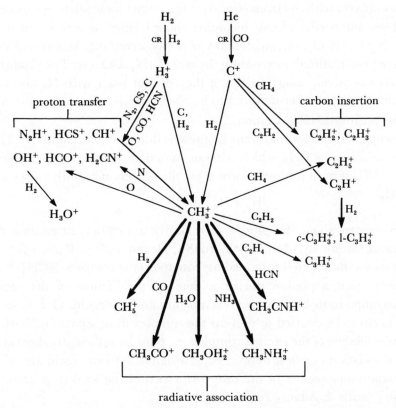

FIGURE 1. The action of cosmic rays (CR) on the abundant H_2 and He creates H_2^+ and He^+ ions, which react with H_2 and CO respectively to generate H_3^+ and C^+. The H_3^+ ions can then proton transfer to a range of atoms and molecules including C, CS and N_2 as shown generating ions such as CH^+, HCS^+, N_2H^+ etc., and the C^+ ions can undergo carbon-insertion reactions with hydrocarbon molecules generating longer-chain hydrocarbon ions. CH_3^+ ions (which are produced from H_3^+ and C^+ as indicated) undergo a variety of reactions including radiative association (indicated by thickened lines) under the low-temperature conditions of interstellar clouds, thus generating large molecular ions in a single step. The rate coefficients and product-ion distributions for many of the reactions included in the figure have been measured by various techniques and are given in the compilations by Albritton (1978) and Anicich & Huntress (1986) and in the reviews by Adams & Smith (1983, 1987 *b*, *c*) that also review the experimental methods. Neutral atoms and molecules are formed from the ions via the process of dissociative recombination (see §3 of text and figure 3).

allowed, that is when the proton affinity (A_p) of a molecule exceeds that of potential protonated donor molecule, then at low temperature proton transfer will occur at every collision between the ion and the molecule (the rate coefficients for these processes are of the order of 10^{-9} cm^3 s^{-1} for non-polar molecules and, as has recently been shown, are as large as 10^{-7} cm^3 s^{-1} for polar molecules at low temperatures (Clary 1985; Adams *et al.* 1985). The relatively small A_p of H$_2$ (4.39 eV) ensures that H$_3^+$ is an excellent proton donor to most molecules (see figure 1). Also, because $A_p(N_2) = 5.13$ eV, N$_2$H$^+$ can transfer a proton to many interstellar species including CO $(A_p = 6.15$ eV), HCN $(A_p = 7.43$ eV) and CS $(A_p = 8.16$ eV). Similarly, HCO$^+$ can proton transfer to HCN and CS generating H$_2$CN$^+$ and HCS$^+$. This well-understood process of proton transfer is extremely important in gas-phase ion chemistry (Adams & Smith 1983) not least in interstellar dense clouds because it not only generates new ions but, in doing so, also creates neutral molecules. For example, in the proton transfer reaction

$$H_3O^+ + NH_3 \rightarrow H_2O + NH_4^+, \tag{4}$$

H$_2$O is known to be formed, whereas this may not necessarily be the case when H$_3$O$^+$ recombines dissociatively with electrons (see §3). Many other molecular ions, as yet undetected, must exist in dense interstellar clouds including those formed as a result of proton transfer reactions of H$_3^+$, N$_2$H$^+$, HCO$^+$, etc., with many of the observed interstellar molecules (listed in table 1) that have proton affinities exceeding those of H$_2$, N$_2$, CO, etc. The product ions of such reactions will have relatively long lifetimes if they do not react with H$_2$ and could then be present in sufficient large number densities to be observed by radioastronomy (assuming that they can radiate via rotational transitions).

A relatively simple ion species that must be present in interstellar clouds is CH$_5^+$. This ion is probably the precursor ion of CH$_4$, which, although not detected because of its symmetry, must surely be present. CH$_5^+$ is most probably formed via the process of radiative association between CH$_3^+$ ions and H$_2$

$$CH_3^+ + H_2 \rightarrow CH_5^+ + h\nu. \tag{5}$$

Much has been written about the process of radiative association in relation to interstellar molecular synthesis (Smith & Adams 1977, 1978a; Herbst 1980b; Bates 1983; Barlow *et al.* 1984). In this process, the excited intermediate ion–molecule complex $((CH_5^+)^*$ in the case of reaction (5)) must emit a photon during a time τ_d, the lifetime of the complex against unimolecular decomposition back to the reactant ion and molecule. Only a low-energy (infrared) photon needs to be emitted to prevent the complex from separating back to reactants. Thus the radiative lifetime of the excited complex, τ_r, must be sufficiently short compared with τ_d to allow the association reaction to proceed at a significant rate. Estimates of τ_d have been obtained from laboratory studies of the rate coefficients of the analogous termolecular association processes (Smith & Adams 1978b), e.g.

$$CH_3^+ + H_2 + He \rightarrow CH_5^+ + He, \tag{6}$$

in which the intermediate complex is stabilized in a superelastic collision with another species that exists in large concentration in the reaction cell (in this case He atoms). Theory predicts (Bates 1979, 1980; Herbst 1979, 1980a, b) and experiments confirm (Meotner 1979; Adams & Smith 1981a, b) that τ_d increases as the temperature of the reactants decreases and, for strongly bound intermediates, τ_d can greatly exceed 10^{-2} s, which is expected to be the upper-limit τ_r for infrared emissions (Herbst 1985b) although it may be much shorter for some polyatomic ions radiatively decaying through vibrational states near to their dissociation limit. Thus radiative association is confidently invoked as an important process for the production of complex molecular ions under the low-temperature conditions of dense interstellar clouds (Herbst *et al.* 1984). This process along with several other processes, is included in the very general scheme shown in figure 1 for the production of interstellar ions from which, ultimately, the observed interstellar neutral molecules are produced. To date, radiative emission has not been observed from ion–molecule association and there is a great need for such spectroscopic studies. More information is also required on the infrared radiative lifetimes of excited complex molecular ions. However, there is little doubt that radiative association is a reality, as the classic study of reaction (5) at low temperatures has shown in which the rate coefficient has been measured as 1.8×10^{-13} cm^3 s^{-1} at 13 K (Barlow *et al.* 1984). This is more than adequate to initiate the production of CH_5^+ from CH_3^+ in dense interstellar clouds. That CH_5^+ does not react with H_2 ensures that it will persist in the clouds. Its ultimate fate will be either to transfer its proton to another species generating CH_4 or to dissociatively recombine with electrons, perhaps again producing CH_4 or CH_3 (see §3). The relative importance of these two destruction processes will depend on the relative number density of the proton-acceptor molecules to that of electrons in the cloud and on the respective rate coefficients for the reactions.

The importance of CH_3^+ in dense interstellar cloud ion chemistry is clear from figure 1. Its relative unreactivity with H_2 means that a small but significant fraction of CH_3^+ ions are available to react with other molecules via the process of radiative association (e.g. with CO, H_2O and HCN) and via normal bimolecular fragmentation reactions such as those with N and O atoms and hydrocarbon molecules. The radiative association reactions generate ions containing N and O as well as C and H, which are the precursor ions to neutral molecules such as CH_3OH and CH_3CN. The reactions of CH_3^+ with N and O atoms are another source of H_2CN^+ and HCO^+ ions (Fehsenfeld 1976) and reactions of CH_3^+ with hydrocarbon molecules containing n carbon atoms generally produce ions with $(n+1)$ carbon atoms (Herbst *et al.* 1983). Such reactions are often termed carbon insertion reactions and, as can be seen from figure 1, reactions of C^+ with hydrocarbon molecules also result in carbon insertion. Generally, such reactions will most probably result in the extension of the carbon chain, but in the rapid reaction

$$CH_3^+ + C_2H_2 \rightarrow c\text{-}C_3H_3^+ + H_2 \tag{7}$$

the lowest energy isomer of $C_3H_3^+$ (the cyclic isomer, c-$C_3H_3^+$) is formed exclusively (Adams & Smith 1987d). However, the major source of interstellar c-$C_3H_3^+$ (which is probably the precursor ion of the recently discovered, yet ubiquitous, interstellar cyclic molecule c-C_3H_2, (Matthews & Irvine 1985; Thaddeus *et al.* 1985)) is probably the sequence of reactions (Herbst *et al.* 1984)

$$C \xrightarrow{C_2H_2} C_3H^+ \xrightarrow{H_2} c\text{-}C_3H_3^+, l\text{-}C_3H_3^+ \xrightarrow{e} c\text{-}C_3H_2, l\text{-}C_3H_2. \tag{8}$$

Note that both the linear and cyclic isomers of $C_3H_3^+$ are formed in the second stage, which is the association reaction of C_3H^+ with H_2. (Both l-$C_3H_3^+$ and c-$C_3H_3^+$ were observed as products of the collisional association reaction of C_3H^+ with H_2 and are presumed also to result from radiative association (Adams & Smith 1987d; Smith & Adams 1987).) Clearly, as the complexity of the ions increases then a greater number of structural isomers is possible. Although few experimental data exist on the structures of the ion products of gas-phase ion–molecule reactions, this is an exciting and a growing area of interest that is obviously very relevant to interstellar chemistry because other cyclic molecules must exist in interstellar clouds. The beautiful experimental techniques (Adams & Smith 1987 b) that have been developed to study ionic reactions in the gas phase and that have provided the kinetic data that substantiate the ion chemistry summarized in figure 1, will soon, no doubt, lead to the identification of other cyclic ions that could be the precursors of other cyclic interstellar molecules.

Several interstellar molecular species are detected also containing the rare isotopes D, ^{13}C, ^{15}N, ^{17}O, ^{18}O, ^{29}Si, ^{33}S and ^{34}S. This is in itself no surprise, but what was initially surprising and very interesting was the observation that the fractional abundances of the molecules containing the rare isotope to those of the same species containing the common isotope, exceeded their terrestrial ratios. This was especially evident for the deuterated species for which the D:H ratio for a given species was greater by one or two orders of magnitude than the terrestrial ratio. The ^{13}C:^{12}C ratio was also greater than the terrestrial value for several species but typically only by about a factor of two. It is now appreciated that these enhanced isotope ratios in interstellar molecules are largely because of the phenomenon of isotopic fractionation in ion–molecule reactions (Watson 1976; Smith et al. 1982 a, b; Smith & Adams 1980, 1984 a; Adams & Smith 1981 a, b, 1985) as exemplified by the elementary reaction

$$D^+ + H_2 \rightleftharpoons H^+ + HD + \Delta E, \quad \Delta E/k = 462 \text{ K}. \tag{9}$$

This reaction is appreciably exothermic to the right ($\Delta E = 39.8$ meV; equivalent to $\Delta E/k = 462$ K, k is the Boltzmann constant) mostly because the vibrational zero-point energy (ZPE) of H_2 exceeds that of HD (Henchman et al. 1981). The result of this is that the reverse of reaction (9) is very slow at interstellar cloud temperatures (i.e. when $T \ll \Delta E/k$) and so the reaction proceeds only in the forward direction. This forward reaction is very efficient, occurring essentially on every collision between a D^+ ion and an H_2 molecule, and therefore it fractionates deuterium into HD. Hence, much of the D in dense interstellar clouds is in the form of HD (together with a fraction of D atoms and smaller fractions of other deuterated molecules). Detailed laboratory studies of deuterium fractionation in ion–molecule reactions have been carried out over a range of temperatures down to 80 K (Smith et al. 1982 a, b; Smith & Adams 1984 a; Adams & Smith 1985). From these studies has emerged a clear idea as to how most of the observed deuterated interstellar molecules are formed. Again, H_3^+ and CH_3^+ are strongly implicated in this, because they both react very rapidly with HD (but not with H_2) by

$$H_3^+ + HD \rightleftharpoons H_2D^+ + H_2, \quad \Delta E/k = 220 \text{ K} \tag{10}$$
$$(\text{at } T = 20 \text{ K}),$$

$$CH_3^+ + HD \rightleftharpoons CH_2D^+ + H_2, \quad \Delta E/k = 335 \text{ K}, \tag{11}$$

thus generating their deuterated analogues H_2D^+ and CH_2D^+. (Note that H_2D^+ has been tentatively identified in interstellar clouds, by Phillips et al. 1985.) The reverse of reactions (10) and (11), which would act to convert the deuterated ions back to H_3^+ and CH_3^+, are very slow at low temperatures (again when $T \ll \Delta E/k$), whereas the forward reactions are rapid and occur essentially on every collision between the ions and an HD molecule. Hence, the H_2D^+ and CH_2D^+ are able to react with other species in the gas clouds (in a totally analogous way to H_3^+ and CH_3^+; see figure 1) generating molecules that are enriched in deuterium. The deuterated molecules observed to date in interstellar clouds together with the elementary reactions (including reactions (9)–(11)) that are considered to generate them are shown in figure 2. Also included in figure 2 is the D–H exchange reaction

$$C_2H_2^+ + HD \rightleftharpoons C_2HD^+ + H_2, \quad \Delta E/k = 550 \text{ K}. \tag{12}$$

By this reaction, deuterium is fractionated into C_2HD^+, which on recombining with an electron can generate the observed interstellar species C_2D. It is observed, somewhat surprisingly, that the $C_2D:C_2H$ ratio is greater in the Orion cloud ($T \approx 50$ K) than in the Taurus cloud, TMC-1 ($T \approx 10$ K) even though laboratory studies have clearly established that deuterium fractionation is more rapid at low temperatures. This has been explained by recognizing that both $C_2H_2^+$ and C_2HD^+ radiatively associate with H_2, which competes more effectively with reaction (12) and with dissociative recombination for the destruction of $C_2H_2^+$ and C_2HD^+ at the lower temperature of TMC-1 and hence diminishes the rate of production of C_2D (Herbst *et al.* 1987). The C_2HD^+ produced in reaction (12) may also be involved in the production of the

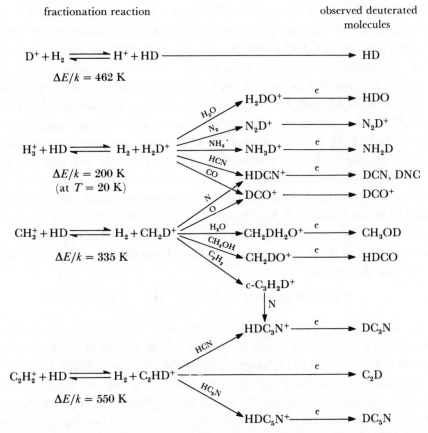

FIGURE 2. The H–D exchange (fractionation) reactions are shown as reversible reactions but because of their large exothermicities, ΔE (expressed as $\Delta E/k$ in Kelvin, where k is the Boltzmann constant), they only proceed to the right in cold dense clouds. Thus HD, H_2D^+, CH_2D^+ and C_2HD^+ are efficiently produced by the reactions indicated and then can react with the molecules shown generating the deuterated ions shown, of which both N_2D^+ and DCO^+ have been detected in dense interstellar clouds. Kinetic data on the H–D exchange reactions are given in the review papers by Adams & Smith (1983) and Smith & Adams (1984 *a*). Source references for the list of deuterated interstellar molecules are given in Winnewisser *et al.* (1979) and Irvine *et al.* (1987). Dissociative recombination reactions of the other deuterated ions with electrons can produce the deuterated neutral molecules indicated, all of which have been detected in dense clouds, although it must be said that the neutral products of these recombination reactions have not yet been established experimentally (see §3 of text and figure 3).

DC_3N and DC_5N detected in interstellar clouds by the reactions indicated in figure 2, although this chemistry is still uncertain.

It has been shown that D–H exchange is very facile at low temperatures especially for 'symmetrical' proton-bound dimer systems including the systems $H_3^+ \cdot H_2$ (i.e. H_2—H^+—H_2, of which reaction (10) is an example), $H_3O^+ \cdot H_2O$, $NH_4^+ \cdot NH_3$ and $CH_5^+ \cdot CH_4$ etc., and this has been explained as being caused by the 'shuttling' of the proton (or deuteron) between the two chemically identical molecules, which are also able to rotate within the (strongly bound) proton-bound dimer (Adams et al. 1982). However, it should be noted that there are many ion–molecule interactions in which D–H exchange does not occur at a measurable rate even at a temperature of 80 K. These include the interactions of the important interstellar ions HCO^+, N_2H^+, H_3O^+, NH_4^+ and CH_5^+ with HD. For such cases, the ion–H_2 complexes are only very weakly bound, the binding energies being much less than the difference between the proton affinities of the proton donating molecules (CO, N_2, H_2O, NH_3 and CH_4) and that of H_2 (which is the smallest). This, in effect, presents an energy barrier to the motion of the proton. Thus, transfer (or shuttling) of the proton within the ion–molecule complexes cannot occur because of the presence of this energy barrier and so H–D exchange is inhibited. That the above-mentioned ions are not deuterated in collisions with HD has helped greatly in understanding and explaining the relative abundances of the deuterated species observed in interstellar clouds. It is interesting to note that, although D–H exchange occurs efficiently in reaction (12) between $C_2H_2^+$ and HD, it does not occur to a measurable degree when $C_2H_3^+$ interacts with HD. This observation has helped in explaining the apparently anomalous difference in the $C_2D:C_2H$ ratio between the Orion cloud and the Taurus cloud discussed above (Herbst et al. 1987). It has also helped in characterizing the vibrational spectra of $C_2H_3^+$ (Oka, personal communication 1986). It is also worthy of note that it has recently been shown that some of those ions that do not isotope exchange with HD do, however, exchange with D atoms (Adams & Smith 1985). This needs to be taken into account when considering D–H exchange in the interstellar clouds (Dalgarno & Lepp 1984).

Isotope fractionation of the heavy isotopes of other elements also occurs in the ion–molecule reactions such as in the reaction

$$^{13}C^+ + {}^{12}CO \rightleftharpoons {}^{12}C^+ + {}^{13}CO, \quad \Delta E/k = 40 \text{ K}. \tag{13}$$

The relatively small $\Delta E/k$ for reaction (13) means that ^{13}C fractionation only occurs to a significant extent at temperatures much lower than those at which deuterium fractionation occurs. Nevertheless, reaction (13) is largely responsible for the enhanced $^{13}CO:^{12}CO$ ratios (relative to the terrestrial $^{13}C:^{12}C$ ratio) in dense interstellar clouds (Smith & Adams 1980). The $\Delta E/k$ for the exchange of ^{14}N for ^{15}N in the interaction of N_2H^+ with N_2 is only about 10 K and thus fractionation in this system can only occur at very low temperatures (Adams & Smith 1981 a, b). The detailed laboratory studies of isotope exchange in ion–molecule reactions and the interstellar implications have been discussed in some recent reviews (Smith 1981, 1987; Smith & Adams 1984a; Wootten 1987).

3. Destruction of positive ions in interstellar clouds: dissociative recombination

Clearly, a given positive-ion species may be destroyed in reaction with neutral molecules as, for example, by those reaction processes that are described above and are illustrated in figure 1. The simplest and fastest of these processes is proton transfer (e.g. reactions (2) and (4)) in which predictable new molecular and ionic species are generated, and this may be the final process in the production of a particular observed interstellar molecule. For example, it has been demonstrated that proton transfer from H_2Cl^+ to CO is most probably the final step in the production of the HCl recently detected in interstellar clouds (Blake *et al.* 1985) that proceeds thus:

$$Cl^+ \xrightarrow{H_2} HCl^+ \xrightarrow{H_2} H_2Cl^+ \xrightarrow{CO} HCO^+ + HCl. \tag{14}$$

The individual reactions included in the reaction sequence (14) have all been studied in laboratory experiments (Smith & Adams 1981 *a*, *b*, 1985) and the HCO^+ and HCl products of the final proton-transfer step are quite certain.

The alternative process for the destruction of positive ions in any ionized gas, including the weakly ionized interstellar clouds, is dissociative recombination. With very few exceptions (notably H_3^+, see below) molecular positive ions are neutralized by electrons and the resulting neutral excited molecule then fragments. This is the process of dissociative recombination. Sufficient energy is usually released to ensure that one or more of the neutral fragments are not only kinetically excited but also internally excited. Indeed, for simple polyatomic ions total fragmentation to atoms may occur when it is energetically allowed. This is possible for the recombination reaction of H_2Cl^+ with electrons that could produce two H atoms and a Cl atom (compare this with the final step in the reaction sequence (14)). Herein is the major problem in invoking dissociative recombination to explain the production of neutral molecules from positive ions in gas-phase ion chemistry (such as that illustrated in figure 2). Little is known about the products of such reactions except for those involving diatomic ions where the products can only be two atoms, and a few triatomic ions where the energetics dictate the nature of the products. Examples of these simple cases are given in figure 3. Clearly, the products of dissociative recombination of O_2^+ ions can only be two O atoms but interest is then directed to determining the ways in which the energy released is divided between translation and internal (electronic) modes of excitation. Experiments have shown that for O_2^+ recombination, the O atom products are distributed among the ^3P ground state and the metastable ^1S and ^1D states (Zipf 1970, 1980), but even this apparently simple case is complicated in that branching into the various states of the O atoms is very dependent on the vibrational state of the recombining O_2^+ ions (Guberman 1983). Almost all that is known about the products of recombination is indicated in the upper half of figure 3 and this has been obtained from emission and absorption spectroscopy of afterglow plasmas. In the most recent of these studies, the products of re-combination of internally excited H_2O^+ ions (for which the internal state of excitation is unknown) have been investigated by detecting the product H and O atoms by vacuum ultraviolet (vuv) absorption (Rowe *et al.* 1987).

Although the data obtained for the products of O_2^+, N_2^+ and NO^+ dissociative recombination (Zipf 1970, 1980; Kley *et al.* 1977; Queffelec *et al.* 1985) are of great interest to terrestrial aeronomy, recombination of these ions with electrons (and also the CO_2^+ and H_2O^+ indicated

$$(O_2^+)^* \xrightarrow{\ e\ } O(^3P, {}^1S, {}^1D) \quad : \quad N_2^+(v = 1) \xrightarrow{\ e\ } N(^2D) + N(^2D)$$

$$(NO^+)^* \xrightarrow{\ e\ } N(^2D) + O(?) \quad : \quad (CO_2^+)^* \xrightarrow{\ e\ } CO(A^1\Pi) + O(?)$$

$$(H_2O^+)^* \xrightarrow{\ e\ } \begin{cases} O + H + H \ (> 30\%) \\ OH + H \quad (\sim 55\%) \\ O + H_2 \quad (< 15\%) \end{cases}$$

What are the products of recombination of polyatomic ground-
state interstellar ions?

Obvious are: $HCO^+ \xrightarrow{\ e\ } H + CO^*$ only energetically
                  $N_2H^+ \xrightarrow{\ e\ } H + N_2^*$ allowed channels.

But what about:

H_3O^+	NH_4^+	CH_5^+	H_2CN^+	$CH_3OH_2^+$	$c\text{-}C_3H_3^+$
$\downarrow e$	$\downarrow e$	$\downarrow e$	$\downarrow e$	$\downarrow e$	$\downarrow e$
H_2O?	NH_3?	CH_4?	HCN?	CH_3OH?	$c\text{-}C_3H_2$?
OH?	NH_2?	CH_3?	CN?	radicals?	$l\text{-}C_3H_2$?

Are H atoms ejected?
Are radicals produced?

FIGURE 3. Little is known about the products of dissociative recombination reactions of interstellar ions with electrons. Some information is available (as indicated in the upper half of the figure) concerning the products of recombination of O_2^+, NO^+, CO_2^+ and H_2O^+ in unknown states of internal excitation and for $N_2^+(v = 1)$, none of which is an important interstellar ion (for references, see §3). The neutral products of HCO^+ and N_2H^+ recombinations are as indicated, these being the only energetically allowed products (the asterisks associated with the various species, including the CO and N_2, indicate that the species are internally excited). From the viewpoint of interstellar chemistry, the products of recombination of several important interstellar ions such as those indicated at the bottom of the figure (H_3O^+, --- $c\text{-}C_3H_3^+$, etc.) are urgently required. It would be particularly interesting to know whether the cyclic or linear (or both) isomers of C_3H_2 are produced when $c\text{-}C_3H_3^+$ recombines with electrons.

in figure 3) does not occur to any significant extent in interstellar clouds because either they are not formed to any significant degree (i.e. O_2^+ and NO^+) or they react much more rapidly with the abundant H_2 than with the much less abundant electrons (i.e. N_2^+, CO_2^+ and H_2O^+). This can be easily demonstrated by calculating the reaction rates by using the measured rate coefficients for the reactions of these ions with H_2 and the measured recombination coefficients. Although little is known about the products of dissociative recombination, a great many experimental data have been obtained over the last three decades concerning the recombination coefficients and their temperature dependences by using stationary afterglow methods (Bardsley & Biondi 1970; Biondi 1973). Quite recently, a flowing afterglow technique has been developed to determine recombination coefficients over a wide range of temperature for a wide variety of polyatomic ions, including many interstellar ions (e.g. H_3O^+, HCO^+ and CH_5^+)

(Smith & Adams 1984b, c). The collected results of all these studies indicate that, generally, recombination coefficients increase with increasing complexity of the recombining ion and with decreasing temperature. Thus, for most interstellar polyatomic molecular ions recombining with electrons at low temperatures, a recombination coefficient of $ca.$ 10^{-6} cm^3 s^{-1} is appropriate. By using this laboratory result together with estimates of the degree of ionization of interstellar clouds, it can be shown that dissociative recombination is the most important loss process for most interstellar ions that do not react rapidly with H_2 and this includes those given in the lower half of figure 3. So, the outstanding question is: 'What are the products of recombination of such ions?' So often in interstellar ion chemical models it is assumed for expediency that an H atom is ejected following recombination leaving a stable molecule. This must indeed by the case for the recombination of HCO^+ and N_2H^+ because this is the only energetically allowed channel in both cases (see figure 3). But this is certainly not the only possibility for most other interstellar ions such as those examples given in figure 3. A statistical theory has been developed to predict the products of recombination of the important interstellar ions CH_3^+, H_3O^+, NH_4^+ and H_2CN^+ (Herbst 1978), which, for example, indicates that $CH + H_2$ will be favoured over $CH_2 + H$ as the major product of CH_3^+ recombination and that OH will be strongly favoured over H_2O in H_3O^+ recombination. However, in a more recent paper (Bates 1986), it is argued that dissociative channels involving the least rearrangement of valence bonds are favoured, and it is concluded that the predictions of the statistical theory are not valid.

Clearly, theoretical calculations of the products of recombination of polyatomic ions are particularly difficult and experiments are necessary to answer the questions posed at the bottom of figure 3. In this regard, spectroscopy on the neutral products offers the best hope of success and collaborative experiments are underway between the Universities of Rennes, France (B. Rowe and J. L. Queffelec), and Birmingham, England (N. G. Adams and D. Smith) that will surely provide the answers to these questions for some ions. These new experiments rely on the fact that H_3^+ ions in the ground vibrational state do not recombine with electrons at a measurable rate (it is interesting to note that the Birmingham group have shown that HeH^+ also does not recombine with electrons). This remarkably important result for interstellar chemistry has only quite recently been demonstrated experimentally (Smith & Adams 1984b) and substantiated by theory (Michels & Hobbs 1984). It has made it necessary to reappraise the ion–chemical models of interstellar chemistry and the method of predicting electron densities in dense clouds (Smith & Adams 1984b). The new laboratory experiments are essentially based on the Birmingham FALP apparatus (Smith & Adams 1984c; Adams & Smith 1987c) to which spectroscopic diagnostics are being added to detect the products of recombination of vibrationally relaxed ions (e.g. H and O atoms by vuv absorption and product molecular fragments, e.g. OH and CN radicals by laser-induced fluorescence). The first stage in the experiment is to create an H_3^+–electron flowing afterglow plasma. This is achieved by adding H_2 in excess to a helium afterglow plasma consisting of largely He_2^+ ions, metastable helium atoms and electrons. Charge exchange between the He_2^+ and H_2 and Penning ionization of H_2 by helium metastables both create H_2^+, which rapidly reacts with H_2 producing H_3^+ and an H atom. The latter is detected by Ly-α absorption and effectively calibrates the experiment. A proton-acceptor molecule, e.g. H_2O, can then be added to the H_3^+ plasma producing (via proton transfer) H_3O^+ ions, which are rapidly recombining species. If on recombination H atoms are generated then these are detected as an increase in the Ly-α absorption. In this way, the number of H atoms produced per recombining ion can be determined. Additionally, if radicals such as OH are generated then these can be detected by laser-induced fluorescence. Very preliminary results on the Ly-α experiment have indicated that valuable data on the products of recombination of several molecular ions (importantly in their ground states) will soon be obtained. Such will be a major contribution to interstellar chemistry and also provide much valuable fundamental data on dissociative recombination. Perhaps it might be possible to answer the intriguing question posed at the bottom of figure 3, which is whether cyclic neutral molecules (e.g. c-C_3H_2) are generated when cyclic ions (e.g. c-$C_3H_3^+$) recombine with electrons!

4. SUMMARY AND CONCLUDING REMARKS

Positive ions are created in interstellar clouds by the action of electromagnetic and cosmic radiation on the ambient gas. The simple (primary) ions thus produced then react with ambient atoms and molecules by a variety of mechanisms generating polyatomic ions (see figure 1). Those ions that are reactive with H_2 are relatively quickly destroyed because of the preponderance of H_2 in interstellar clouds, but those that are unreactive with H_2 have a longer lifetime. Significantly, as laboratory work has shown, those ions that have been detected in dense interstellar clouds do not react with H_2 at significant rates even at low temperatures. It is clear from the ion chemistry outlined in this paper that many more ionic species must exist in interstellar clouds than have so far been detected. Other species will undoubtedly be detected as their spectra are determined and as the detection sensitivities of radio antennae are increased. Those molecular ions that are unreactive with ambient neutrals are eventually destroyed via the process of dissociative recombination with electrons. This is probably the final process by which most of the neutral molecules detected in interstellar clouds are formed (see table 1). Dissociative recombination is generally an efficient reaction process but because the ionization density in dense interstellar clouds is low ($[e]/[H_2] \approx 10^{-7}$) (Smith & Adams 1984$a$–$c$; Dalgarno & Lepp 1984), then the number densities of several ionic species should be large enough for them to be detected, notably those ions that are formed efficiently and that do not react with H_2. The recent recognition that ground-state H_3^+ recombines only slowly, if at all, (Smith & Adams 1984b, 1987) with electrons means that its reactions with neutrals to generate other ion species takes on a greater significance than had previously been thought. Also, the recent recognition that its proton-transfer reactions with polar molecules are much more rapid at low temperatures than had previously been assumed (Adams et $al.$ 1985) means that ions that essentially are protonated interstellar polar molecules, are prime candidates for detection. Significantly, H_2CN^+ has recently been detected in concentrations consistent with ion-chemical predictions based on the new realizations concerning the reactions of H_3^+ ions (Millar et $al.$ 1985; Ziurys & Turner 1986). It is well worth noting that this understanding of interstellar ion chemistry can be put to good use in creating laboratory plasmas consisting of particular ion species and in facilitating the generation of ion beams for spectroscopic studies. Hopefully, it will also stimulate spectroscopists to select for study some of those ions that are identified as being involved in interstellar chemistry.

Isotope exchange in ion–molecule reactions, particular H–D exchange, is a very interesting phenomenon both fundamentally and from the viewpoint of interstellar chemistry. Laboratory studies have shown that several of the important ions involved in interstellar molecular synthesis, including H_3^+ and CH_3^+, undergo H–D exchange in reaction with HD. This process is very efficient at low temperatures and is responsible for the fractionation of the heavy isotopes of some elements into interstellar ions and neutral molecules. All of the deuterated molecules so far detected in interstellar clouds (see figure 2) are enriched in deuterium to varying degrees and, significantly, H_2D^+ has recently been (tentatively) detected in interstellar clouds. However, laboratory studies have also shown that some important interstellar ions do not D–H exchange with HD (including HCO^+, H_3O^+, NH_4^+ and $C_2H_3^+$) and this is valuable additional information in the efforts to elucidate the synthetic routes to interstellar molecules. Again it can be a help to spectroscopists in identifying the spectra of certain ions.

It is clear that molecular spectroscopy has a vital role to play in the further study of molecular ions in the gas phase, and thus in the field of interstellar chemistry. The accurate characterization of the rotational and vibrational spectra of more ion species is required before they can be usefully sought in interstellar clouds. It seems likely that structural isomers of some ions (e.g. $C_3H_3^+$) must exist in interstellar clouds and, if the spectra of these could be separately characterized in laboratory experiments, it would be a great step forward. It has been postulated (§2) that radiative association is a major reaction process in the synthesis of interstellar molecules, yet a major uncertainty in quantitatively accounting for this process in interstellar ion–chemical models lies in the magnitudes of the radiative lifetimes of excited ion complexes. Spectroscopic work is urgently required as a guide in this area. Indeed, it would be of great value if the structures of the ions formed in certain ion–molecule reactions, including some association reactions, could be ascertained since this is also a major area of doubt. It has already been stressed how important it is to the proper understanding of interstellar molecular synthesis to be able to specify the neutral products of dissociative recombination of polyatomic positive ions. Again in this, as in so much of the interstellar research, spectroscopic studies are the best and often the only hope of gaining the required understanding. Interstellar physics and chemistry is both a mission field and an exciting stimulus to spectroscopy.

I am indebted to my colleague Dr N. G. Adams for his invaluable assistance in the preparation of this paper.

References

Adams, N. G. & Smith, D. 1981 a *Astrophys. J.* **247**, L123.
Adams, N. G. & Smith, D. 1981 b *Chem. Phys. Lett.* **79**, 563.
Adams, N. G. & Smith, D. 1983 In *Reactions of small transient species* (ed. A. Fontijn & M. A. A. Clyne), pp. 311–385. London: Academic Press.
Adams, N. G. & Smith, D. 1985 *Astrophys. J.* **294**, L63.
Adams, N. G. & Smith, D. 1987 a In *Astrochemistry* (ed. M. S. Vardya & S. P. Tarafdar), pp. 1–18. Dordrecht: Reidel.
Adams, N. G. & Smith, D. 1987 b *Science Progr.* **71**, 91.
Adams, N. G. & Smith, D. 1987 c In *Techniques of chemistry* (ed. J. M. Farrar & W. H. Saunders, Jr). New York: Wiley–Interscience. (In the press.)
Adams, N. G. & Smith, D. 1987 d *Astrophys. J.* **317**, L25.
Adams, N. G., Smith, D. & Henchman, M. J. 1982 *Int. J. Mass Spectrom. Ion Phy.* **42**, 11.
Adams, N. G., Smith, D. & Clary, D. C. 1985 *Astrophys. J.* **296**, L31.
Albritton, D. L. 1978 *Atom. Data Nucl. Data Tables* **22**, 1.
Anicich, V. G. & Huntress, W. T. Jr 1986 *Astrophys. J. Suppl. Ser.* **62**, 553.
Bardsley, J. N. & Biondi, M. A. 1970 *Adv. atom. molec. Phys.* **6**, 1.
Barlow, S. E., Dunn, G. H. & Schauer, M. 1984 *Phys. Rev. Lett.* **52**, 902.
Bates, D. R. 1979 *J. Phys.* B **12**, 4135.
Bates, D. R. 1980 *J. chem. Phys.* **73**, 1000.
Bates, D. R. 1983 *Astrophys. J.* **270**, 564.

Bates, D. R. 1986 *Astrophys. J.* **306**, L45.

Biondi, M. A. 1973 *Comments atom. molec. Phys.* **4**, 85.

Blake, G. A., Keene, J. & Phillips, T. G. 1985 *Astrophys. J.* **295**, 501.

Bohme, D. K., Mackay, G. I. & Schiff, H. I. 1980 *J. chem. Phys.* **73**, 4976.

Bohme, D. K. 1975 In *Interactions between ions and molecules* (ed. P. Ausloos), pp. 489–504. New York: Plenum Press.

Churchwell, E., Woods, R. C., Dickman, R. L. & Irvine, W. M. 1987 (In preparation.)

Clary, D. C. 1985 *Molec. Phys.* **54**, 605.

Crutcher, R. M. & Watson, W. D. 1985 In *Molecular astrophysics: state of the art and future directions* (ed. G. H. F. Diercksen, W. F. Huebner & P. W. Langhoff), pp. 255–280. Dordrecht: Reidel.

Dalgarno, A. & Black, J. H. 1976 *Rep. Prog. Phys.* **39**, 573.

Dalgarno, A. & Lepp, S. 1984 *Astrophys. J.* **287**, L47.

Douglas, A. E. & Herzberg, G. 1941 *Astrophys. J.* **94**, 381.

Douglas, A. E. & Herzberg, G. 1942 *Can. J. Res.* A**20**, 71.

Duley, W. W. & Williams, D. A. 1984 *Interstellar chemistry*, pp. 86–114. London: Academic Press.

Dunham, T. Jr & Adams, W. S. 1937*a* *Bull. Am. astr. Soc.* **9**, 5.

Dunham, T. Jr & Adams, W. S. 1937*b* *Bull. Astr. Soc. Pac.* **49**, 26.

Fehsenfeld, F. C. 1976 *Astrophys. J.* **209**, 638.

Guberman, S. L. 1983 In *Physics of ion–ion and electron–ion collisions* (ed. F. Brouillard & J. W. McGowan), pp. 167–200. New York: Plenum.

Guelin, M. 1985 In *Molecular astrophysics: state of the art and future directions* (ed. G. H. F. Diercksen, W. F. Huebner & P. W. Langhoff), pp. 23–43. Dordrecht: Reidel.

Guelin, M. 1987 In *Astrochemistry* (ed. M. S. Vardya & S. P. Tarafdar), pp. 171–181. Dordrecht: Reidel.

Henchman, M. J., Adams, N. G. & Smith, D. 1981 *J. chem. Phys.* **75**, 1201.

Herbst, E. 1978 *Astrophys. J.* **222**, 508.

Herbst, E. 1979 *J. chem. Phys.* **70**, 2201.

Herbst, E. 1980*a* *J. chem. Phys.* **72**, 5284.

Herbst, E. 1980*b* *Astrophys. J.* **241**, 197.

Herbst, E. 1985*a* In *Molecular astrophysics: state of the art and future directions* (ed. G. H. F. Diercksen, W. F. Huebner & P. W. Langhoff), pp. 237–254. Dordrecht: Reidel.

Herbst, E. 1985*b* *Astrophys. J.* **291**, 226.

Herbst, E. 1987 In *Astrochemistry* (ed. M. S. Vardya & S. P. Tarafdar), pp. 235–244. Dordecht: Reidel.

Herbst, E., Adams, N. G. & Smith, D. 1983 *Astrophys. J.* **269**, 329.

Herbst, E., Adams, N. G. & Smith, D. 1984 *Astrophys. J.* **285**, 618.

Herbst, E., Adams, N. G., Smith, D. & DeFrees, D. J. 1987 *Astrophys. J.* **312**, 351.

Hollis, J. M. & Rhodes, P. J. 1982 *Astrophys. J.* **262**, L1.

Huntress, W. T. Jr & Anicich, V. G. 1976 *Astrophys. J.* **208**, 237.

Irvine, W. M. 1987 In *Astrochemistry* (ed. M. S. Vardya & S. P. Tarafdar), pp. 245–252. Dordrecht: Reidel.

Irvine, W. M., Goldsmith, P. F. & Hjalmarson, A. 1987 In *Summer school on interstellar processes* (ed. D. J. Hollenbach & H. A. Thronson). Dordrecht: Reidel.

Kim, J. K., Theard, L. P. & Huntress, W. T. Jr 1974 *Int. J. Mass. Spectrom. Ion Phys.* **15**, 223.

Kley, D., Lawrence, G. M. & Stone, E. J. 1977 *J. chem. Phys.* **66**, 4157.

Lias, S. G., Liebman, J. F. & Levin, R. D. 1984 *J. Phys. Chem. Ref. Data* **13**, 695.

Leung, C. M., Herbst, E. & Huebner, W. F. 1984 *Astrophys. J. Suppl. Ser.* **56**, 231.

Matthews, H. E. & Irvine, W. M. 1985 *Astrophys. J.* **298**, L61.

Meotner, M. M. 1979 In *Gas phase ion chemistry*, vol. 1 (ed. M. T. Bowers), pp. 197–271. New York: Academic Press.

Michels, H. H. & Hobbs, R. H. 1984 *Astrophys. J.* **286**, L27.

Millar, T. J. 1982 *Mon. Not. R. astr. Soc.* **199**, 309.

Millar, T. J. 1985 In *Molecular astrophysics: state of the art and future directions* (ed. G. H. F. Diercksen, W. F. Huebner & P. W. Langhoff), pp. 613–619. Dordrecht: Reidel.

Millar, T. J. & Freeman, A. 1984*a* *Mon. Not. R. astr. Soc.* **207**, 405.

Millar, T. J. & Freeman, A. 1984*b* *Mon. Not. R. astr. Soc.* **207**, 425.

Millar, T. J. & Nejad, L. A. M. 1985 *Mon. Not. R. astr. Soc.* **217**, 507.

Millar, T. J., Adams, N. G., Smith, D. & Clary, D. C. 1985 *Mon. Not. R. astr. Soc.* **216**, 1025.

Millar, T. J., Leung, C. M. & Herbst, E. 1987 *Astron. Astrophys.* **183**, 109.

Oka, T. 1981 *Phil. Trans. R. Soc. Lond.* A**303**, 543.

Phillips, T. G., Blake, G. A., Keene, J., Woods, R. C. & Churchwell, E. 1985 *Astrophys. J.* **294**, L45.

Queffelec, J. L., Rowe, B. R., Morlais, M., Gomet, J. C. & Vallee, F. 1985 *Planet. Space Sci.* **33**, 263.

Rowe, B. R., Vallee, F., Queffelec, J. L., Gomet, J. C. & Morlais, M. 1987 (Pre-print.)

Rydbeck, O. E. H. & Hjalmarson, A. 1985 In *Molecular astrophysics: state of the art and future directions* (ed. G. H. F. Diercksen, W. F. Huebner & P. W. Langhoff), pp. 45–175. Dordrecht: Reidel.

Smith, D. 1981 *Phil. Trans R. Soc. Lond.* A**303**, 535.

Smith, D. 1987 *Phil. Trans. R. Soc. Lond.* A**323**, 269.

Smith, D. & Adams, N. G. 1977 *Astrophys. J.* **217**, 741.

Smith, D. & Adams, N. G. 1978*a* *Astrophys. J.* **220**, L87.

Smith, D. & Adams, N. G. 1978*b* *Chem. Phys. Lett.* **54**, 535.

Smith, D. & Adams, N. G. 1980 *Astrophys. J.* **242**, 424.

Smith, D. & Adams, N. G. 1981*a* *Int. Rev. Phys. Chem.* **1**, 271.

Smith, D. & Adams, N. G. 1981*b* *Mon. Not. R. astr. Soc.* **197**, 377.

Smith, D. & Adams, N. G. 1984*a* In *Ionic processes in the gas phase* (ed. M. A. Almoster-Ferreira), pp. 41–66. Dordrecht: Reidel.

Smith, D. & Adams, N. G. 1984*b* *Astrophys. J.* **284**, L13.

Smith, D. & Adams, N. G. 1984*c* In *Swarms of ions and electrons in gases* (ed. W. Lindinger, T. D. Mark & F. Howorka), pp. 284–306. Vienna: Springer-Verlag.

Smith, D. & Adams, N. G. 1985 *Astrophys. J.* **298**, 827.

Smith, D. & Adams, N. G. 1987 *Int. J. Mass Spectrom. Ion Processes* **76**, 307.

Smith, D., Adams, N. G. & Alge, E. 1982*a* *J. chem. Phys.* **77**, 1261.

Smith, D., Adams, N. G. & Alge, E. 1982*b* *Astrophys. J.* **263**, 123.

Thaddeus, P., Guelin, M. & Linke, R. A. 1981 *Astrophys. J.* **246**, L41.

Thaddeus, P., Vrtilek, J. M. & Gottlieb, C. A. 1985 *Astrophys. J.* **299**, L63.

van Dishoeck, E. F. 1987 In *Astrochemistry* (ed. M. S. Vardya & S. P. Tarafdar), pp. 51–65. Dordrecht: Reidel.

Vardya, M. S. & Tarafdar, S. P. (eds) 1987 *Astrochemistry*. Dordrecht: Reidel.

Watson, W. D. 1976 *Rev. mod. Phys.* **48**, 513.

Watson, W. D. 1977 In *Topics in interstellar matter* (ed. H. van Woerden), pp. 135–147. Dordrecht: Reidel.

Williams, D. A. 1987 In *Astrochemistry* (ed. M. S. Vardya & S. P. Tarafdar), pp. 531–538. Dordrecht: Reidel.

Winnewisser, G. & Herbst, E. 1987 *Topics in current chemistry* vol. 139, pp. 119–172. Berlin: Springer-Verlag.

Winnewisser, G., Churchwell, E. & Walmsley, C. M. 1979 In *Modern aspects of microwave spectroscopy* (ed. G. W. Chantry), pp. 313–501. New York: Academic Press.

Winnewisser, M., Winnewisser, B. P. & Winnewisser, G. 1985 In *Molecular astrophysics: state of the art and future directions* (ed. G. H. F. Diercksen, W. F. Huebner & P. W. Langhoff), pp. 375–402. Dordrecht: Reidel.

Woods, R. C. 1987 In *Astrochemistry* (ed. M. S. Vardya & S. P. Tarafdar), pp. 77–85. Dordrecht: Reidel.

Woods, R. C., Gudeman, C. S., Dickman, R. L., Goldsmith, P. F., Huguenin, G. R. & Irvine, W. M. 1983 *Astrophys. J.* **270**, 583.

Wootten, A. 1987 In *Astrochemistry* (ed. M. S. Vardya & S. P. Tarafdar), pp. 311–319. Dordrecht: Reidel.

Zipf, E. C. 1970 *Bull. Am. phys. Soc.* **24**, 129.

Zipf, E. C. 1980 *J. geophys. Res.* **85**, 4232.

Ziurys, L. M. & Turner, B. E. 1986 *Astrophys J.* **302**, L31.

Phil. Trans. R. Soc. Lond. A **324**, 275–287 (1988)

Printed in Great Britain

Spectroscopy of the hydrogen molecular ion at its dissociation limit

BY A. CARRINGTON†, F.R.S., I. R. McNAB AND CHRISTINE A. MONTGOMERIE

Physical Chemistry Laboratory, University of Oxford, South Parks Road, Oxford OX1 3QZ, U.K.

Previous theoretical and experimental work on the ion H_2^+ and its deuterium isotopes HD^+ and D_2^+ is briefly reviewed. Spectroscopic studies of the vibration–rotation levels are discussed, and recent work on the infrared photodissociation of the high-lying vibrational levels is described. Earlier work that made use of ion-beam techniques to study the vibration–rotation levels of HD^+ is reviewed, and compared with the most recent theoretical predictions.

The nuclear-hyperfine and spin–rotation structure of HD^+ is described, and recent observations of the vibration–rotation satellite lines that will yield absolute values of the deuterium hyperfine constants are presented. We conclude by describing our attempts to observe radiofrequency–infrared double-resonance spectra.

1. INTRODUCTION

The hydrogen molecular ion occupies a central position in the development of molecular quantum mechanics. A detailed review (Carrington & Kennedy 1984) has been presented elsewhere, and a short summary of the underlying theory here will suffice to establish the context of our current work. The hydrogen molecular ion (and its isotopic variants) contains two nuclei and one electron, and is therefore the molecular example of the quantum-mechanical three-body problem. Because this problem cannot be solved exactly, theoretical studies have been directed towards appropriate approximate treatments that yield values of the energy levels and other molecular constants. Electron–electron interactions are, of course, not present in this simple molecule; consequently the focus of attention is on the nuclear and electronic motions, and their mutual interactions. Calculations may be classified under three main headings. Born–Oppenheimer calculations attempt to solve the problem of the electron moving in the electrostatic field of two fixed nuclei. The Schrödinger equation for this system may be separated into three one-dimensional differential equations (Alexandrow 1926) and exact numerical solutions obtained (Burrau 1927). Analytical solutions with series expansions were developed (Hylleraas 1931; Jaffé 1934) and have been the basis of much subsequent work.

Inclusion of cross terms between the nuclear and electronic kinetic-energy operators that are diagonal in the ground electronic state constitutes the adiabatic approximation and the first complete calculations of all bound vibration–rotation levels of H_2^+, HD^+ and D_2^+ were provided by Hunter *et al.* (1974). Figure 1 shows the highest-lying vibrational levels of the three isotopes in their ground electronic state ($^2\Sigma_g^+$); the first excited electronic state ($^2\Sigma_u^+$), which also correlates with the lowest dissociation limits, is repulsive at all internuclear distances (except for a van der Waals minimum), a fact that is crucial to our experimental studies. Non-adiabatic corrections to these energy levels involve the coupling of different electronic states by the nuclear kinetic energy. This problem was first tackled by Bishop & Cheung (1977), and calculations for all bound vibrational levels have been presented by Wolniewicz & Poll (1978, 1986). Radiative

† Present address: Department of Chemistry, University of Southampton, Southampton SO9 5NH, U.K.

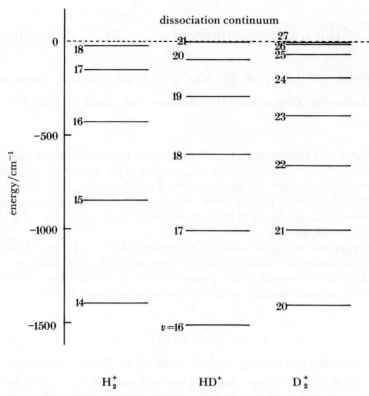

FIGURE 1. Highest-energy vibrational levels for H_2^+, HD^+ and D_2^+ in their ground electronic states, based upon the calculations of Hunter *et al.* (1974).

and relativistic corrections have also been included in the calculations of Wolniewicz & Poll.

Laboratory spectroscopic studies of H_2^+ and its isotopes are difficult, particularly because the only bound excited electronic states predicted by theory lie at energies more than 11 eV above the ground state; conventional electronic spectroscopy is therefore not possible. A notable early achievement was the measurement of the radiofrequency spectra of H_2^+ in vibrational levels $v = 4$–8 by an ingenious method that involved quadrupole trapping of the ion and spatial alignment produced by photodissociation with polarized white light (Dehmelt & Jefferts 1962; Jefferts 1968, 1969). The principles underlying this method will be discussed later. Apart from this exceptional study, however, most of our earlier knowledge of the vibrational levels of H_2^+ and its isotopes came from photoelectron studies (Åsbrink 1970; Pollard *et al.* 1982), photoionization studies (Chupka & Berkowitz 1969; Peatman 1976) and Rydberg spectra (Takezawa 1970 a, b; Herzberg & Jungen 1972). The first measurement of a vibration–rotation spectrum of HD^+ (and indeed, of any molecular ion) was described in 1976 by Wing *et al.* They employed Doppler tuning of a fast HD^+ ion beam to bring vibration–rotation transitions into resonance with a carbon monoxide infrared laser frequency; excitation was detected through the resulting changes in the charge-exchange attenuation of the HD^+ beam, produced by added neutral gases. Spectra involving the lowest vibrational levels ($v = 0$–3) were studied, and transition frequencies accurate to 0.001 cm^{-1} determined.

2. Infrared photodissociation

An important feature of ion-beam studies is that electron impact ionization of H_2 and its isotopes results in population of all the bound vibrational levels of the ground state up to the dissociation limit; these populations are preserved in the collision-free environment of an ion beam system. Because the first excited state is repulsive, ions in the highest-energy vibrational levels can be photodissociated with infrared radiation. This point was appreciated by Fournier et al. (1979) who calculated cross sections for photodissociation in the frequency range covered by the carbon dioxide laser (900–1100 cm^{-1}). The photodissociation is central to our spectroscopic studies of HD^+, and we have recently carried out a crossed-beam study in which the relative photodissociation cross sections for the highest-energy vibrational levels of H_2^+, HD^+ and D_2^+ are essentially determined. The apparatus, which is also used for the spectroscopic studies described in §3, has been described extensively elsewhere (Carrington & Buttenshaw 1981; Carrington & Kennedy 1984) and is illustrated schematically in figure 2. The ions are formed by electron-impact ionization, with electron energies up to 100 eV, and are accelerated to potentials up to a maximum of 10 kV. Mass analysis of the ion beam is achieved with a 55° magnetic sector, and the desired ion beam is then brought into coincidence with a focused carbon dioxide infrared laser beam, aligned either collinear with, or perpendicular to, the ion beam. Photofragment ions are separated from the parent ions with an 81.5° electrostatic analyser (ESA) and detected with an off-axis electron multiplier.

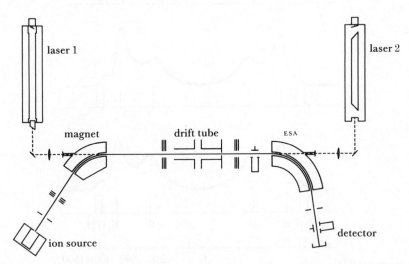

FIGURE 2. Schematic diagram of the tandem mass spectrometer (Vacuum Generators ZAB 1F instrument with modifications).

The ion beam comes to a focal point that is at the centre of a cylindrical drift tube situated between the magnetic and electric sectors. This region of the apparatus is very important for the spectroscopic studies described in the next section, and will be discussed in more detail therein.

The energy resolution of the ESA is sufficiently high for the laboratory kinetic energy of the photofragment ions to be determined; a suitable transformation to the centre of mass can then be performed. Because the electronic transition in H_2^+ leading to photodissociation is polarized along the internuclear axis, crossed-beam irradiation, with the electric vector parallel to the ion beam, leads to preferential production of protons in the forward and backward directions of the parent ion beam. The irradiation point is close to the ion-beam focal point at the centre of the drift tube, and the photofragment kinetic-energy spectra are obtained by scanning the ESA voltage. Slits before and after the ESA are adjusted to optimize resolution, at the cost of fragment ion-beam intensity. Our best results for H_2^+, which were obtained by chopping the laser beam and detecting the AC photofragment current with a lock-in amplifier, are shown in figure 3;

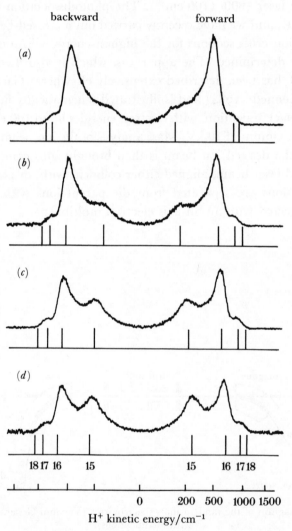

FIGURE 3. Photofragment H^+ kinetic-energy spectra obtained for H_2^+ at four different laser frequencies: (a) 938.69; (b) 979.70; (c) 1041.28; (d) 1082.29 cm^{-1}.

spectra at four different laser wavelengths with an average laser power of 30 W cw are shown. Vibrational structure is resolved, and it is clear that the photodissociation cross section is largest for $v = 16$; it should be noted, however, that the spectra shown in figure 3 reflect both the photodissociation cross sections, the vibrational-level populations, and instrumental factors that give an intensity dependence roughly proportional to the inverse of the centre-of-mass kinetic-energy release. Structure due to ions in the highest vibrational level, $v = 18$, is not resolved and the cross section at these frequencies is probably small, but crucially not zero. Similar studies of HD^+ and D_2^+ show only one peak for HD^+ (from $v = 18$) at all four laser frequencies, but with D_2^+ peaks due to ions in $v = 22$, 23 and 24 are resolved, particularly at the lowest-used laser frequency of 938.69 cm^{-1}.

3. Two-photon vibration–rotation spectra of HD^+

Figure 4 shows the potential curves for the ground and first excited electronic states of HD^+, and the highest vibrational levels of the ground state. As we have already noted, ions in $v \geqslant 18$ will be photodissociated by infrared radiation from a carbon dioxide laser (900–1100 cm^{-1}) but those of low rotational quantum number in $v \leqslant 17$ will not. Consequently vibration–rotation transitions, such as those belonging to the 17–20 band shown in figure 4, can be detected through the resulting increase in the number of photoproduct H^+ or D^+ ions. Most of our previous work has been carried out with the apparatus shown in figure 2. The laser beam is aligned to be collinear with the ion beam, either parallel or antiparallel. Coarse frequency scanning is achieved by selecting an appropriate carbon dioxide laser line, and fine tuning results from scanning the Doppler shift by changing the ion-beam potential.

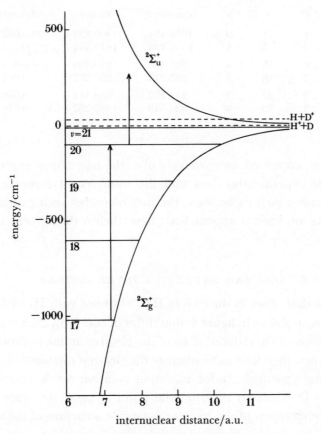

FIGURE 4. Potential-energy curves for the ground and first excited electronic states of HD^+, in the vicinity of the dissociation limits. (1 a.u. (length) = 1 Bohr $\approx 0.529 \times 10^{-10}$ m.)

For an ion as light as HD^+, beam potentials from 1 to 10 kV, with positive or negative Doppler shifts, provide complete frequency tuning from one laser line to the next.

The ESA is used to separate the fragment ions (we usually choose to monitor D^+) from the parent ions. We apply an oscillating voltage to the drift tube shown in figure 2; this produces velocity modulation of the ion beam and hence, through the Doppler effect, frequency modulation. Consequently, vibration–rotation transitions are located by detecting the AC photofragment current from the electron multiplier with a lock-in amplifier.

The sensitivity of the method is high, and transitions in the $v = 16$–18 band an be detected with signal:noise ratios of approximately 1000:1 (Carrington & Buttenshaw 1981). It proved to be advantageous to use two lasers, one operating at high powers (typically 20 W cw) to maximize the photofragment intensity, and the other at lower powers (1–2 W) to drive the vibration–rotation transitions. Measurements of rotational components of the 16–18, 14–17, 15–17 and 17–20 bands have been described. In many cases we were able to make more than one independent measurement of a particular vibration–rotation transition by using different laser lines at different ion-beam potentials, aligned parallel or antiparallel. The resulting cancellation of systematic errors means that vibration–rotation frequencies were measured with an accuracy better than 0.001 cm^{-1}. This provides a severe test of the accuracy of *ab initio*

TABLE 1. VIBRATION–ROTATION TRANSITIONS (RECIPROCAL CENTIMETRES) IN HD^+

v''	N''	v'	N'	experiment	theory	difference
0	1	1	0	1869.134	1869.131	$+0.003$
2	2	3	1	1642.108	1642.107	$+0.001$
16	0	18	1	926.490	926.494	-0.004
16	2	18	3	933.213	933.217	-0.004
17	0	20	1	918.102	918.111	-0.009
17	3	20	4	885.749	885.767	-0.018

calculations, which have improved progressively over the past fifteen years. Table 1 presents a brief selection of the experimental data and the most recent theoretical calculations of Wolniewicz & Poll (1986). As can be seen, the difference between experiment and theory increases as the dissociation limit is approached; nevertheless the agreement is impressive.

4. NUCLEAR HYPERFINE INTERACTIONS

A fascinating feature that arises in the case of HD^+ (but not with H_2^+ or D_2^+) is the presence of two dissociation limits, as shown in figure 4, that differ in energy by 29.8 cm^{-1}. This difference arises because of the difference in reduced mass of the electron in the hydrogen and deuterium atoms. The question arises, therefore, as to whether the electron distribution in HD^+ is sensitive to this energy difference, particularly for vibration–rotation levels approaching the lower dissociation limit, $H^+ + D$. From an experimental point of view, the route to answering this question lies through observation of the nuclear hyperfine structure of the vibration–rotation transitions.

The proton and deuteron have nuclear spins of $\frac{1}{2}$ and 1 respectively, and there will be magnetic hyperfine interactions with the electron spin S of $\frac{1}{2}$. In addition, one expects a magnetic interaction between the magnetic moments caused by electron spin and nuclear rotation. The expected relative magnitudes of these interactions suggests that the most appropriate scheme for coupling the various angular momenta is the following:

$$S + I_H = G_1 \quad G_1 = \quad 1 \quad ; 0,$$
$$G_1 + I_D = G_2 \quad G_2 = 2, 1, 0; 1,$$
$$G_2 + N = F.$$

Possible values of the quantum numbers G_1 and G_2 are shown. The most significant hyperfine constants, with approximate estimates of their values for the highest-lying vibrational levels, are as follows:

b_H, the proton Fermi contact interaction (less than or equal to 720 MHz);
b_D, the deuteron Fermi contact interaction (more than or equal to 110 MHz);
t_H, the axial component of the proton dipolar interaction (less than or equal to 12 MHz);
t_D, the axial component of the deuteron dipolar interaction (less than or equal to 2 MHz);
γ, the spin–rotation interaction (less than or equal to 10 MHz).

Many other interactions are also present, but their magnitudes will be much smaller. Carrington & Kennedy (1985) have given explicit expressions for all the relevant matrix elements of the hyperfine interactions in the above basis, and have also made theoretical calculations of the values of different hyperfine constants. The hyperfine level structure for the vibration–rotation transition 16, 1–18, 2 is shown in figure 5. For each vibration–rotation level there is a large doublet splitting ($G_1 = 1$ and 0) arising from the proton Fermi contact interaction, a triplet splitting ($G_2 = 2, 1, 0$) of the $G_1 = 1$ level, from the deuteron Fermi contact term, and much smaller splittings of each G_1, G_2 level caused by the spin–rotation interaction.

It is clear from figure 5 that many different transitions are possible, and we shall discuss all of these in due course. However, the allowed electric-dipole vibration–rotation transitions must satisfy the selection rules

$$\Delta N = \pm 1, \quad \Delta G_1 = 0, \quad \Delta G_2 = 0, \quad \Delta F = 0, \pm 1.$$

Leaving aside the small spin–rotation splittings, we see that there are four transitions that satisfy these selection rules. Any observed hyperfine splitting of the allowed vibration–rotation transitions will depend only on the differences in hyperfine splitting for the upper and lower states.

At high laser powers, the observed vibration–rotation resonances consist of a single line of width 30–40 MHz, but at lower laser powers it was soon found that every transition exhibits a characteristic doublet splitting (typically 10–20 MHz). The two lines arise from levels with $G_1 = 0$ or 1 and the splitting shows that the proton hyperfine splitting in the upper level is the smaller.

For vibration–rotation levels approaching the dissociation limit $H^+ + D$, one might expect the proton splitting in HD^+ to decrease, with a corresponding increase in the deuteron splitting. The deuteron magnetogyric ratio is nearly seven times smaller than that of the proton, so that changes in the deuteron hyperfine splitting are much smaller and therefore more difficult to observe. Observation of levels with $v = 20$ or 21 would provide a critical test of these expectations and they were indeed confirmed by Carrington & Kennedy (1985) through

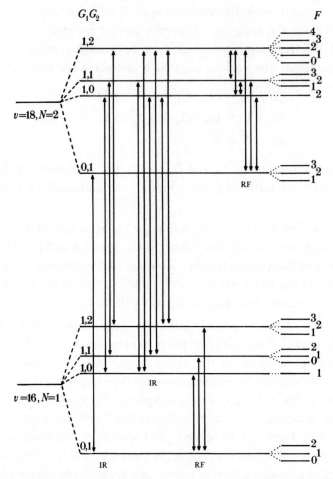

FIGURE 5. Nuclear-hyperfine and spin–rotation structure of two vibration–rotation levels of HD⁺. The electric-dipole allowed vibration–rotation transitions are shown on the left, the forbidden transitions are shown in the centre, and possible radiofrequency transitions are shown to the right.

measurements of the 17–20 band. The observed proton hyperfine splittings ranged from 80 to 125 MHz, and additional splitting due to the increased deuteron interaction was observed, proving unambiguously that the electron distribution in HD⁺ does indeed become asymmetric as the H⁺ + D dissociation limit is approached. The highest level to be observed was $v = 20$, $N = 4$, which lies 47 cm⁻¹ below the dissociation limit. Theoretical calculations that took into account the non-adiabatic coupling of the ground and first excited states provided a satisfactory interpretation of the observed spectra.

It is unfortunate that the allowed electric-dipole transitions are necessarily diagonal in G_1 and G_2 and hence cannot usually yield absolute values of the nuclear hyperfine constants. Transitions that formally follow selection rules $\Delta G_1 = \pm 1$ or $\Delta G_2 = \pm 1$ acquire electric-dipole intensity only through the spin–rotation and nuclear-dipolar interactions, which mix states differing in G_1 and G_2. The mixing is predicted to be very small, because the necessary interactions are small, so that $\Delta G_2 \neq 0$ transitions such as those shown in figure 5 are predicted to be three orders of magnitude weaker than the fully allowed transitions. Weak satellites were reported by Carrington & Buttenshaw (1981) but they were not reproducible and were not studied in detail.

[208]

We have recently built a new ion-beam machine that has resulted in increased sensitivity, so that the $\Delta G_2 \neq 0$ satellites are clearly and reproducibly observable. The new machine is shown schematically in figure 6, and is very similar to an earlier instrument (Carrington *et al.* 1979). It consists of two 90° magnetic sectors, with focusing lenses between them, and an ion-beam path between the sectors of only 50 cm. Ironically, this machine was actually built for

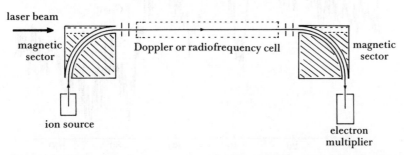

FIGURE 6. Tandem ion-beam system employing two 90° magnetic sectors (see text).

the double-resonance experiments described in the next section. It is much smaller than the machine described in figure 2; consequently tighter focusing of the ion and laser beams is possible, and this probably accounts for the increased sensitivity.

Figure 7 shows a very recent recording of the 16, 1–18, 2 transition in which the $\Delta G_2 = \pm 1$ satellites are clearly observable. Guided by the calculations of Carrington & Kennedy (1985) we are able to assign these satellites, as shown. Similar satellite structure has been observed for the 16, 2–18, 3 transition and we expect to be able to determine absolute values of at least the deuteron Fermi contact constants for the upper and lower states. The enhanced sensitivity has also enabled us to improve the resolution of the strongly allowed electric–dipole transitions, and figure 8 shows the 16, 1–18, 2 transition, now split into four components, rather than just two as described recently. We can assign the values of G_1 with some confidence; it is tempting

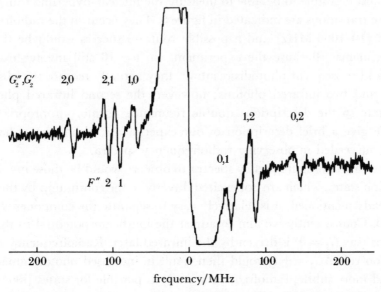

FIGURE 7. The 16, 1–18, 2 vibration–rotation transition (932.2237 cm^{-1}) in HD$^+$, showing the $\Delta G_2 = \pm 1$ satellites.

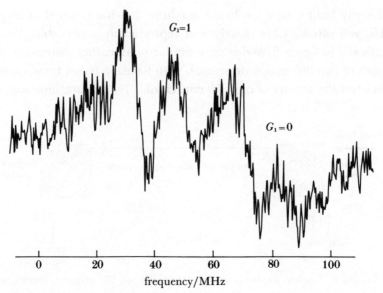

FIGURE 8. The resolution of the electric-dipole allowed components of the 16, 1–18, 2 vibration–rotation transition (932.2237 cm^{-1}) in HD$^+$, plotted on a relative frequency scale in MHz.

to assign the three $G_1 = 1$ peaks to $G_2 = 2, 1, 0$, but this may be an over-simplification. There is every reason to suppose that we will be able to observe $\Delta G_2 = \pm 1$ satellite transitions for other vibrational-band systems.

Unfortunately, the $\Delta G_1 = \pm 1$ satellite transitions, which would yield the proton hyperfine constants, are predicted to be very much weaker than the $\Delta G_2 = \pm 1$ transitions because the hyperfine states are much more widely separated and therefore less mixed by the coupling terms. We have not yet succeeded in observing such satellites, but the searches will continue.

5. RADIOFREQUENCY–INFRARED DOUBLE-RESONANCE EXPERIMENTS

It would be most desirable to be able to measure the nuclear-hyperfine transitions directly; examples of these transitions are indicated in figure 5. They occur in the radiofrequency region of the spectrum (10–1000 MHz) and a possible route to success would be through double-resonance experiments. (Because the experiments on $v = 16$ still involve excitation to the repulsive state and subsequent photodissociation, they actually require the absorption of one radiofrequency and two infrared photons; however, the second infrared photon excites to a continuum state so the description 'double resonance' is more appropriate than 'triple resonance'.) We give a brief description of our experiments, although we have not, at the time of writing, succeeded in observing radiofrequency spectra.

Potentially the easiest radiofrequency spectra to observe would be those involving the lower vibration–rotation states, which are not excited directly to the continuum by the infrared laser. As we have already mentioned, it is relatively easy to separate the components corresponding to $G_1 = 1$ and 0. Consequently we aim to adjust the ion-beam potential so that one of these transition groups (say $G_1 = 0$) is driven by the infrared laser. Radiofrequency transitions that obey the selection rule $\Delta G_1 = \pm 1$ would then result in increased photofragmention.

A second, and more subtle, radiofrequency study is possible for states (like $v = 18$, $N = 2$, shown in figure 5) that are excited directly to the repulsive state by the infrared laser. The principles involved are exactly those described and exploited by Dehmelt & Jefferts (1962) in their study of the lower vibrational levels of H_2^+. Photodissociation of 18, 2, for example, obeys the selection $\Delta M_F = 0$, where the quantization axis is defined by the direction of the electric vector of the infrared laser. Consequently states of different M_F value photodissociate at different rates, so that the remaining undissociated ions exhibit a degree of spatial alignment. The development of spatial alignment is reflected in the cross section for photodissociation. Pumping of different F levels by a radiofrequency magnetic field will change the degree of alignment, and for $\Delta M_F = \pm 1$ transitions remove it altogether, so that the photodissociation rate will increase. We therefore monitor the production of photofragment ions as a function of the radiofrequency. It is not necessary to drive simultaneously an allowed vibration–rotation transition. One of the most attractive features of this experiment is that, because magnetic dipole transitions are involved, we are not restricted to the heteronuclear species HD^+.

We have attempted many different methods of coupling radiofrequency power to the ion beam, and our present method is based on that described by Rosner *et al.* (1978) in their studies of the Xe^+ ion. It is shown schematically in figure 9. The radiofrequency cell consists of a

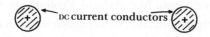

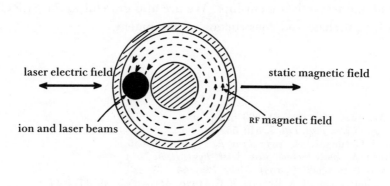

FIGURE 9. Relative orientations of the ion and laser beams, radiofrequency coaxial cell, and four-pole current conductors employed in double-resonance experiments.

coaxial arrangement of inner and outer copper tubes, with the ion and laser beams transmitted between the conductors as shown. The present device is 36 cm in length, and radiofrequency power is coupled in and out by modified coaxial cable connectors. Care is taken to maintain a 50 Ω impedance throughout the frequency range, and we are able to couple in at least 10 W through the range 10–1000 MHz. As shown in figure 9 the radiofrequency magnetic field is mainly perpendicular to the ion-beam axis and to the electric vector of the laser beam. In addition, the coaxial cell resides inside a four-pole structure supplied with DC current to produce a static magnetic field (if required) that is parallel to the laser electric vector. We amplitude-

or frequency-modulate the radiofrequency power, and detect the photofragment current at the modulation frequency.

We have adopted the present method only recently, and have not yet succeeded in detecting radiofrequency resonances. The radiofrequency magnetic field at the ion beam is calculated to be approximately 0.1 G†, and although we believe this should be sufficient, it may be marginal. Resonant high-Q devices would, of course, give much higher radiofrequency fields, but frequency tuning then becomes difficult.

As we have mentioned, one of the most attractive features of the radiofrequency experiment that depends upon spatial alignment is its potential application to the homonuclear species H_2^+ and D_2^+. Our infrared studies have necessarily been restricted to the heteronuclear species, HD^+, which possesses the necessary electric-dipole moment. It may well be the most interesting of the three species because of the existence of the two close-lying dissociation limits, and there is still much to be revealed about the details of this system. Even though the large asymmetry observed for HD^+ is not expected for H_2^+ and D_2^+, it is nevertheless possible that they will also exhibit surprising nuclear-hyperfine effects at their dissociation limits.

We are most grateful to Mr Roger Bowler for his superb workmanship in the extensive machining necessary for this work. We would like to thank Professor S. D. Rosner and Professor R. A. Holt of the University of Western Ontario for their advice concerning the design of the coaxial radiofrequency cell. A.C. thanks the Royal Society for a Research Professorship, and I.R.M. and C.A.M. thank the SERC and The British Petroleum Company plc respectively for post-graduate research studentships. We are also grateful to the SERC for their financial support in the purchase and construction of equipment.

REFERENCES

Alexandrow, W. 1926 *Ann. Phys.* **81**, 603–614.
Åsbrink, L. 1970 *Chem. Phys. Lett.* **7**, 549–552.
Bishop, D. M. & Cheung, L. M. 1977 *Phys. Rev.* A **16**, 640–645.
Burrau, Ø. 1927 *K. danske Vidensk. Selsk. Math. Fys. Meddr.* **7**, 14.
Carrington, A. & Buttenshaw, J. 1981 *Molec. Phys.* **44**, 267–285.
Carrington, A., Buttenshaw, J. & Roberts, P. G. 1979 *Molec. Phys.* **38**, 1711–1715.
Carrington, A. & Kennedy, R. A. 1984 In *Gas phase ion chemistry*, vol. 3 (*Ions and light*) (ed. M. T. Bowers), pp. 393–442. London: Academic Press, Inc.
Carrington, A. & Kennedy, R. A. 1985 *Molec. Phys.* **56**, 935–975.
Chupka, W. A. & Berkowitz, J. 1969 *J. chem. Phys.* **51**, 4244–4268.
Dehmelt, H. G. & Jefferts, K. B. 1962 *Phys. Rev.* **125**, 1318–1322.
Fournier, P., Lassier-Govers, B. & Comtet, G. 1979 In *Laser-induced processes in molecules* (ed. K. L. Kompa & S. D. Smith), pp. 247–251. Berlin & New York: Springer Verlag.
Herzberg, G. & Jungen, Ch. 1972 *J. molec. Spectrosc.* **41**, 425–486.
Hunter, G., Yau, A. W. & Pritchard, H. O. 1974 *Atom. Data Nucl. Data Tables* **14**, 11–20.
Hylleraas, E. A. 1931 *Z. Phys.* **71**, 739–763.
Jaffé, G. 1934 *Z. Phys.* **87**, 535–544.
Jefferts, K. B. 1968 *Phys. Rev. Lett.* **20**, 39–41.
Jefferts, K. B. 1969 *Phys. Rev. Lett.* **23**, 1476–1478.
Peatman, W. B. 1976 *J. chem. Phys.* **64**, 4093–4099.
Pollard, J. E., Trevor, D. J., Reutt, J. E., Lee, Y. T. & Shirley, D. A. 1982 *J. chem. Phys.* **77**, 34–46.
Rosner, S. D., Gaily, T. D. & Holt, R. A. 1978 *Phys. Rev. Lett.* **40**, 851–854.
Takezawa, S. 1970a *J. chem. Phys.* **52**, 2575–2590.

† 1 G = 10^{-4} T.

Takezawa, S. 1970b *J. chem. Phys.* **52**, 5793–5799.
Wing, W. H., Ruff, G. A., Lamb, W. E. & Spezeski, J. J. 1976 *Phys. Rev. Lett.* **36**, 1488–1491.
Wolniewicz, L. & Poll, J. D. 1978 *J. molec. Spectrosc.* **72**, 264–274.
Wolniewicz, L. & Poll, J. D. 1986 *Molec. Phys.* **59**, 953–964.

Discussion

C. WHITHAM (*Department of Chemistry, University of Nottingham, Nottingham, U.K.*). The first excited state of HD^+ is calculated to possess a shallow ion-induced dipole well. Does Professor Carrington think that there is a possibility of observing levels supported by this well?

A. CARRINGTON, F.R.S. The first excited electronic states of H_2^+, HD^+ and D_2^+ each possess long-range minima arising from the charge-induced dipole interaction; these should perhaps be called Langevin minima. They certainly support vibration–rotation levels, and electronic transitions from the ground states are allowed. The only problem is to devise a scheme for detecting such transitions! We are following various possibilities in laboratory work at the present time.

W. H. WING (*Department of Physics, University of Arizona, Tucson, U.S.A.*). I should like to mention two effects in addition to accurate non-adiabatic energies that must be included to gain good agreement between theory and experiment for HD^+. The first of these is the relativistic mass increase of the electron that arises on account of its kinetic energy. The second is the shift in the electron's energy that stems from the zitterbewegung in the electron's motion, caused by quantum-electrodynamic zero-point radiation field fluctuations. This 'jiggling motion' introduces an effective term in the electron hamiltonian that is proportional to the potential curvature near the nuclei and corresponds to the Lamb Shift in atomic energies. When these electronic eigenenergy shifts are averaged over the nuclear motion, they cause spectroscopically observable shifts of about 10 and 5 p.p.m. (parts per million), respectively, in the vibrational–rotational lines. The hydrogen molecular ion is the only molecule for which theory and experiment have both been done accurately enough to require inclusion of these effects, as Professor Carrington's data demonstrate.

Phil. Trans. R. Soc. Lond. A **324**, 289–294 (1988)

Printed in Great Britain

General discussion: electronic emission spectroscopy of CF_4^+ and SiF_4^+

By S. M. Mason and R. P. Tuckett

Department of Chemistry, University of Birmingham, P.O. Box 363, Birmingham B15 2TT, U.K.

with an appendix by R. N. Dixon

We report the observation of electronic emission spectra in the tetrahedral molecular ions CF_4^+ and SiF_4^+. The spectra are observed at a low rotational temperature (less than 30 K) in a crossed molecular-beam – electron-beam apparatus (Carrington & Tuckett 1980). These spectra are especially interesting because the fluorescing states in the two ions lie up to 10 eV above their lowest dissociation channel (to $CF_3^+/SiF_3^+ + F$; see figure 1), and these states might be expected to decay non-radiatively rather than by a radiative channel. The observation of fluorescence decay from highly excited electronic states of these polyatomic ions is therefore a very surprising phenomenon.

Both continuous and discrete bands have been observed for both ions in the visible–ultraviolet region of the electromagnetic spectrum. From photoelectron spectroscopy, the ground and first two excited electronic states ($\tilde{X}$, $\tilde{A}$ and $\tilde{B}$) of CF_4^+ and SiF_4^+ are known to dissociate rapidly, and the continuous bands arise from transitions to these states. These three electronic states arise from electron removal from molecular orbitals in CF_4/SiF_4, which are essentially F $2p\pi$ non-bonding in character. The third and fourth excited electronic states ($\tilde{C}$ and $\tilde{D}$) give vibrational structure in their photoelectron spectra, and hence are bound (figure 1). They arise from electron removal from t_2 and a_1 molecular orbitals in CF_4/SiF_4 that are essentially σ bonding in character. These are the upper states of the bound–free continuous transitions observed in both ions, e.g. CF_4^+ $\tilde{D}$–$\tilde{A}$ at 160 nm, $\tilde{D}$–$\tilde{B}$ at 189 nm, $\tilde{C}$–$\tilde{X}$ at 230 nm (Aarts *et al.* 1987), SiF_4^+ $\tilde{D}$–$\tilde{A}$ at 304 nm (Mason & Tuckett 1987a). The discrete bands arise from an allowed transition between the two bound states $\tilde{D}^2A_1$–$\tilde{C}^2T_2$, and this paper describes this spectrum in CF_4^+ and SiF_4^+ in detail.

Figure 2 shows the discrete band system between 360 and 420 nm obtained with (a) electron impact ionization of CF_4 in a molecular beam ($T_{rot} = 25$ K), (b) He$^+$ impact ionization of CF_4 at room temperature (Aarts 1985). The narrowing of the bands at the lower rotational temperature supports the assignment of the emitter as the parent molecular ion of CF_4, and comparison with photoelectron data suggests that the spectrum is caused by vibronic bands of CF_4^+ $\tilde{D}^2A_1$–$\tilde{C}^2T_2$. A progression is observed only in the ν_1 (a_1) totally symmetric C–F stretching vibration. The absence of any bands involving the ν_2'' (e), ν_3'' (t_2) or ν_4'' (t_2) Jahn–Teller (J.T.) active vibrations shows that there is no J.T. distortion of the $\tilde{C}^2T_2$ state from tetrahedral geometry.

Figure 3a shows a high-resolution spectrum of the 1_2^0 band at 381 nm obtained in the molecular-beam experiment at 25 K. The spectrum shows a 2:1 doublet that is the spin–orbit splitting of $\tilde{C}^2T_2$. Because the higher-wavelength component is the more intense, the $G_{\frac{3}{2}}$ (degeneracy = 4) spin–orbit component of $\tilde{C}^2T_2$ has higher energy than the $E_{\frac{5}{2}}$ (degeneracy = 2) component, and hence the spin–orbit splitting is positive. This is a surprising result because

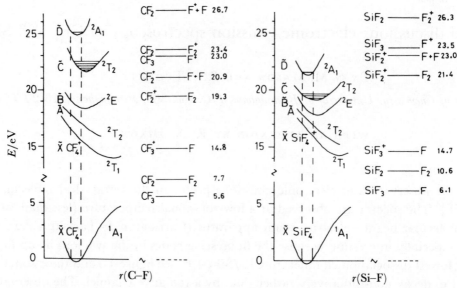

FIGURE 1. Valence ionic states of CF_4 and SiF_4. The energies of the dissociation channels are given in electronvolts.

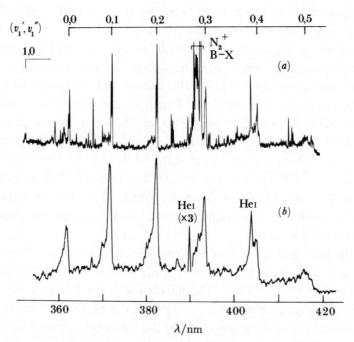

FIGURE 2. (a) The spectrum of CF_4^+ $\tilde{D}-\tilde{C}$ resulting from electron-impact ionization of CF_4 at a low rotational temperature; (b) the same spectrum resulting from He^+ impact (1 keV) excitation at a rotational temperature of *ca.* 300 K. The vibrational assignment (v_1', v_1'') is shown.

this state has the $(t_2)^5$ electron configuration, the t_2 orbital is more than half full, and ξ should be negative. An explanation is provided by Professor R. N. Dixon, F.R.S. (Bristol University) in an Appendix to this paper. Figure 3b shows a simulation of the band with a model developed by Watson for the rotational structure of a $^2A_1-^2T_2$ electronic transition (Watson 1984). The model allows for Coriolis splitting, spin–orbit splitting and J.T. distortion in the $\tilde{C}^2T_2$ state, and

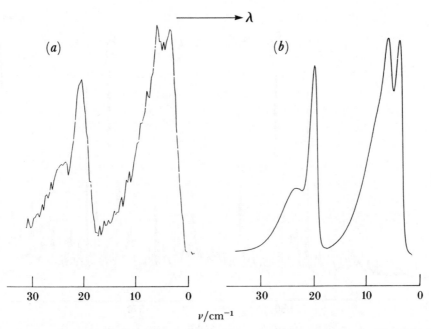

FIGURE 3. (a) The 1_2^0 band of $CF_4^+\tilde{D}-\tilde{C}$ at 381 nm recorded at a resolution of 1.5 cm^{-1}; (b) shows the simulated spectrum.

can be applied to $\tilde{D}-\tilde{C}$ vibronic bands of CF_4^+ or SiF_4^+ that do not involve the J.T. active vibrations (Mason & Tuckett 1987b). The main features of the band are determined by two rotational constants, the Coriolis and spin–orbit constant in $\tilde{C}^2T_2$, and the rotational temperature. Estimates of the two rotational constants are made from the intensity distribution of the vibrational bands in the $\tilde{C}$ and $\tilde{D}$ state photoelectron spectra of CF_4. The agreement with experiment is most satisfactory. The positive sign of the spin–orbit splitting is confirmed, and the Coriolis constant in $\tilde{C}^2T_2$ is substantially reduced from its limiting value of $+1$.

Figure 4 shows the analogous $\tilde{D}-\tilde{C}$ spectrum in SiF_4^+ around 550 nm recorded at a low rotational temperature in the molecular-beam apparatus (Mason & Tuckett 1987a). The spectrum is much more complicated than CF_4^+ $\tilde{D}-\tilde{C}$. The allowed progression in ν_1 is observed, but bands involving ν_2'' and ν_4'' are also seen, showing that the $\tilde{C}^2T_2$ state is distorting from T_d symmetry by J.T. distortion. The strong intensity of 4_1^0 suggests that the lower symmetry of $\tilde{C}^2T_2$ is C_{3v}, and the distortion is probably dynamic in nature.

Figure 5 shows high-resolution spectra of the 0_0^0 and the 4_1^0 J.T.-active band at 551 and 564 nm respectively. This is one of the very few observations of a J.T. distortion in a tetrahedral molecule at 'high' resolution (less than 1 cm^{-1}). The spin–orbit splitting of $\tilde{C}^2T_2$ is resolved in both bands, the sign is now uncertain, and its magnitude is quenched in the 4_1^0 band. Our model of the rotational structure can only be applied to the 0_0^0 band, and the simulation of this band is much more difficult than in CF_4^+ for several reasons. Firstly, estimates of the two rotational constants are less accurate than in CF_4^+. Secondly, ΔB is very small and all the constants become highly correlated. Thirdly, the J.T. effect in $\tilde{C}^2T_2$ means that more constants are needed to determine the rotational structure. However, the gross features of the band are reproduced in the simulation (Mason & Tuckett 1987b), and as in CF_4^+ the sign of the spin–orbit splitting in SiF_4^+ $\tilde{C}^2T_2$ is confirmed to be positive.

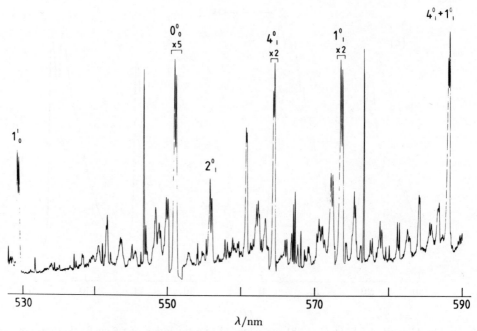

FIGURE 4. The spectrum of SiF_4^+ $\tilde{D}$–$\tilde{C}$ recorded at a low rotational temperature. The resolution is 2.6 cm^{-1}. The very narrow atomic lines are from atomic Si or F.

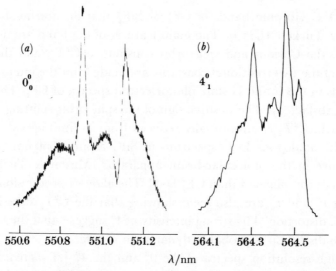

FIGURE 5. (a) The 0_0^0 band of $SiF_4^+\tilde{D}$–$\tilde{C}$ at 551 nm recorded at a resolution of 0.3 cm^{-1}; (b) the 4_1^0 Jahn–Teller band at 564 nm at a resolution of 0.7 cm^{-1}.

In future experiments we wish to obtain fully rotationally resolved spectra of CF_4^+ and $SiF_4^+\tilde{D}$–$\tilde{C}$ at Doppler-limited resolution by photographic techniques. Secondly, we wish to understand the dynamics of the decay pathways of the $\tilde{C}$ and $\tilde{D}$ states of CF_4^+/SiF_4^+. Measurements of fluorescence quantum yields and radiative lifetimes will be made by using synchrotron radiation to ionize CF_4/SiF_4 selectively into these excited electronic states of CF_4^+/SiF_4^+. Thirdly, we wish to understand why $SiF_4^+\tilde{C}^2T_2$ exhibits dynamic J.T. distortion, whereas $CF_4^+\tilde{C}$ does not.

[218]

REFERENCES

Aarts, J. F. M. 1985 *Chem. Phys. Lett.* **114**, 114.
Aarts, J. F. M., Mason, S. M. & Tuckett, R. P. 1987 *Molec. Phys.* **60**, 761.
Carrington, A. & Tuckett, R. P. 1980 *Chem. Phys. Lett.* **74**, 19.
Mason, S. M. & Tuckett, R. P. 1987*a* *Molec. Phys.* **60**, 771.
Mason, S. M. & Tuckett, R. P. 1987*b* *Molec. Phys.* (In the press.)
Watson, J. K. G. 1984 *J. molec. Spectrosc.* **107**, 124.

APPENDIX. THE SIGN OF THE SPIN–ORBIT COUPLING CONSTANT IN $(t_2)^{5\,2}T_2$ STATES OF AX_4^+ IONS

BY R. N. DIXON, F.R.S.

School of Chemistry, University of Bristol, Cantock's Close, Bristol BS8 1 TS, U.K.

The spin–orbit splitting in orbitally degenerate systems with one open shell usually conforms to Hund's third rule: that is, the splitting is regular for a less than half-filled shell, and inverted for a more than half-filled shell. The $\tilde{C}\,^2T_2$ states of CF_4^+ and SiF_4^+ reported by Mason & Tuckett violate this expectation. We show below how this may occur in AX_4^+ ions where the dominant atomic spin-orbit coupling arises from motion around the X atoms.

For $(t_2)^{5\,2}T_2$ states we consider only the spin–orbit coupling matrix elements within the t_2 shell, and use a basis of symmetry-adapted components which transform as translations parallel to the three cubic axes (Dixon *et al.* 1971). There are three types of t_2 basis function involving atomic p-orbitals, one localized on A and two on X_4:

$$\left.\begin{aligned} 1t_{2x} &= x_A, \\ 1t_{2y} &= y_A, \\ 1t_{2z} &= z_A. \end{aligned}\right\} \tag{A 1}$$

$$\left.\begin{aligned} 2t_{2x} &= \tfrac{1}{2}(x_1+x_2+x_3+x_4), \\ 2t_{2y} &= \tfrac{1}{2}(y_1+y_2+y_3+y_4), \\ 2t_{2z} &= \tfrac{1}{2}(z_1+z_2+z_3+z_4). \end{aligned}\right\} \tag{A 2}$$

$$\left.\begin{aligned} 3t_{2x} &= \tfrac{1}{2\sqrt{2}}(y_1-y_2+y_3-y_4+z_1+z_2-z_3-z_4), \\ 3t_{2y} &= \tfrac{1}{2\sqrt{2}}(x_1-x_2+x_3-x_4+z_1-z_2-z_3+z_4), \\ 3t_{2z} &= \tfrac{1}{2\sqrt{2}}(x_1+x_2-x_3-x_4+y_1-y_2-y_3+y_4). \end{aligned}\right\} \tag{A 3}$$

Note that for $3t_2$ the p-orbitals at each atomic centre can be combined into a single p-orbital directed at $90°$ to the translation axis of the basis function (see figure A 1), whereas it is parallel to this axis for $1t_2$ and $2t_2$.

In calculating spin–orbit matrix elements we shall only include integrals involving two atomic orbitals of the same atom. It is then sufficient to use a pseudo one-electron hamiltonian of the following form

$$H'_{S\text{-}O} = \sum_{i=1}^{5} \sum_{k} \xi_k\, l_k(i) \cdot s(i), \tag{A 4}$$

[219]

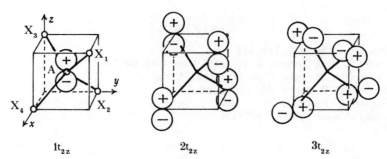

$1t_{2z}$ $2t_{2z}$ $3t_{2z}$

FIGURE A 1. Symmetry-adapted p-orbital basis functions that transform as a translation parallel
to the cubic z-axis.

where l_k is the one-electron angular momentum operator with origin at nucleus k, and ξ_k is the appropriate (positive) atomic-orbital coupling parameter. Let a molecular orbital be expanded as

$$\psi = c_1|1t_2\rangle + c_2|2t_2\rangle + c_3|3t_2\rangle. \tag{A 5}$$

It may then be shown that the splitting of the $^2T_2$ state into its $G_{\frac{3}{2}}$ and $E_{\frac{5}{2}}$ components is isomorphic with that for the effective hamiltonian $\xi_0 \Lambda \cdot S$, where

$$\xi_0 = -[c_1^2 \xi_A + (c_2^2 - \tfrac{1}{2}c_3^2)\,\xi_X]. \tag{A 6}$$

The global negative sign arises from the summation over five electrons, in accordance with Hund's third rule, whereas the additional negative sign to the $3t_2$ contribution is directly related to the atomic orbital orientation commented on above. With $\xi_X > \xi_A$ the limits are $\xi_0 = -\xi_X$ ($c_2 = 1$) and $\xi_0 = +\tfrac{1}{2}\xi_X$ ($c_3 = 1$).

If we assume $p\sigma/p\pi$ separability at the X atoms with respect to the AX bonds, the open shell for the $\tilde{C}^2T_2$ states of CF_4^+ or SiF_4^+ involves the $p\sigma$ combination of $2t_2$ and $3t_2$, with $c_3 = \sqrt{2}\,c_2$. In that approximation there would be no diagonal spin–orbit contribution from the X atoms, and Hund's third rule would be obeyed for the A atom contribution. However, an *ab initio* SCF–MO calculation at the double-zeta level of accuracy for the ground state of neutral CF_4 shows substantial $p\sigma/p\pi$ mixing (Snyder & Basch 1972), with small positive coefficients (c_2) for both basis functions of type $2t_2$, and much larger negative coefficients (c_3) for those of type $3t_2$. There may be some reorganization of the orbitals on ionization to CF_4^+. Nevertheless, it is clear that the multiplier of ξ_F within the bracket in (A 6) must be negative. The overall fluorine contribution to ξ_0 is therefore positive in the $\tilde{C}^2T_2$ state, and will outweigh the negative carbon contribution given that $\xi_F = 269$ cm^{-1} and $\xi_C = 29$ cm^{-1} (2p orbitals). A very similar reasoning applies to the Coriolis constant ζ, except that there are no scaling parameters ξ_F and ξ_C that give heavier weight to the fluorine orbital contribution. The carbon contribution will therefore be positive, but partly offset by a negative fluorine contribution, substantially reducing ζ from its limiting value of $+1$.

In conclusion, we should remember that Hund's rules were formulated for the lowest energy configuration of atoms with one open shell. Their extension to polyatomic molecules may go beyond the range of their validity.

REFERENCES

Dixon, R. N., Murrell, J. N. & Narayan, B. 1971 *Molec. Phys.* **20**, 611.
Snyder, L. C. & Basch, H. 1972 *Molecular wavefunctions and properties*, p. 362. New York: Wiley and Son.